Rethinking Whitehead's *Symbolism*

Thought, Language, Culture

Edited by Roland Faber, Jeffrey A. Bell and Joseph Petek

Edinburgh University Press is one of the leading university presses in the UK. We publish academic books and journals in our selected subject areas across the humanities and social sciences, combining cutting-edge scholarship with high editorial and production values to produce academic works of lasting importance. For more information visit our website: edinburghuniversitypress.com

© editorial matter and organisation Roland Faber, Jeffrey A. Bell and Joseph Petek, 2017
© the chapters their several authors, 2017

Edinburgh University Press Ltd
The Tun – Holyrood Road, 12(2f) Jackson's Entry, Edinburgh EH8 8PJ

Typeset in Sabon
by R. J. Footring Ltd, Derby, UK

A CIP record for this book is available from the British Library

ISBN 978 1 4744 2956 6 (hardback)
ISBN 978 1 4744 2958 0 (webready PDF)
ISBN 978 1 4744 2959 7 (epub)

The right of Roland Faber, Jeffrey A. Bell and Joseph Petek to be identified as the editors of this work has been asserted in accordance with the Copyright, Designs and Patents Act 1988, and the Copyright and Related Rights Regulations 2003 (SI No. 2498).

Contents

Abbreviations	v
Introduction *Joseph Petek*	1

Part I. Perception and Paradox

1 Whitehead on Causality and Perception *Steven Shaviro*	13
2 Originary Symbolism: Whitehead, Deleuze and the Process View on Perception *Keith Robinson*	29
3 Uniting Earth to the Blue of Heaven Above: Strange Attractors in Whitehead's *Symbolism* *Roland Faber*	56

Part II. Adventures in Culture and Value

4 The Inhumanity of Symbolism *Michael Halewood*	81
5 Reverence, Revision and Creaturely Life: Whitehead's Political Theology of Enjoyment *Beatrice Marovich*	96
6 *Ren* and Causal Efficacy: Confucians and Whitehead on the Social Role of Symbolism *Hyo-Dong Lee*	108
7 Avoiding a Fatal Error: Extending Whitehead's Symbolism Beyond Language *Sheri D. Kling*	124

Part III. Misplaced Concreteness in Ethics and Science

8 A Dog's Life: Thought, Symbols and Concepts   147
   Jeffrey Bell

9 From Manipulation to Co-creation: Whitehead on the
   Ethics of Symbol-Making   170
   Luke B. Higgins

10 On Symbols, Propositions and Idiocies: Towards a Slow
   Technoscience   186
   A. J. Nocek

11 Of Symbolism: Climate Concreteness, Causal Efficacy and
   the Whiteheadian Cosmopolis   208
   Catherine Keller

Notes on Contributors   227

Index   231

Abbreviations

Alfred North Whitehead

AI	*Adventures of Ideas* (1933)
CN	*The Concept of Nature* (1920)
ESP	*Essays in Science and Philosophy* (1948)
FR	*The Function of Reason* (1929)
MT	*Modes of Thought* (1938)
PNK	*An Enquiry Concerning the Principles of Natural Knowledge* (1919)
PR	*Process and Reality: An Essay in Cosmology* (1929)
R	*The Principle of Relativity* (1922)
RM	*Religion in the Making* (1926)
S	*Symbolism: Its Meaning and Effect* (1927)
SMW	*Science and the Modern World* (1925)

These abbreviations refer to works by Whitehead and not to any particular published edition. While there are several editions that share common pagination, there are some whose pagination differs between publishers. To find a specific reference, consult the relevant bibliographic list for the chapter in which the reference appears.

Other Abbreviations

ATP	Gilles Deleuze with Félix Guattari, *A Thousand Plateaus* (London: Athlone Press, 1988).
DR	Deleuze, Gilles, *Difference and Repetition* (London: Athlone Press, 1994).
FT	Connolly, William E., *The Fragility of Things: Self-Organizing Processes, Neoliberal Fantasies, and Democratic Activism* (Durham, NC: Duke University Press, 2013).

ML   Williams, Raymond, *Marxism and Literature* (Oxford: Oxford University Press, 1977).
NP   *Nietzsche and Philosophy* (New York: Columbia University Press, 1983).
TF   Deleuze, Gilles, *The Fold: Leibniz and the Baroque* (London: Athlone Press, 1993).

Introduction

JOSEPH PETEK

Alfred North Whitehead's somewhat obscure book *Symbolism: Its Meaning and Effect* was originally delivered in April 1927 as the Barbour-Page lectures, at the University of Virginia, about a year before he gave the Gifford lectures, which would become *Process and Reality*. The latter work often overshadows the former, particularly given that it picks up many of *Symbolism*'s threads while also contextualising them within a larger metaphysical scheme. Yet *Symbolism*'s rather concentrated focus on perception and language is a large part of what recommends the book to our attention, for Whitehead is usually considered a difficult philosopher to grasp, an unsurprising assessment for a thinker whose motto was to 'seek simplicity and distrust it' (CN 163). This is not to say that *Symbolism* is not rich with broader metaphysical and social implications, but only that it is one of his most approachable and accessible works.

Missing Prehensions

Some of the elements that are missing from the book are almost as important as what it contains. For one, it is his only philosophical work which makes no explicit reference to God or religion.[1] For another, Whitehead had not yet decided upon the most suitable terms to express his ideas. It is perhaps not overly surprising that *Symbolism* contains no reference to 'eternal objects' – a concept which Whitehead struggled to communicate clearly, and which he felt was frequently misunderstood by his critics[2] – but even seasoned Whitehead scholars may be surprised to be reminded that the word 'prehension' is nowhere in evidence. Given that 'prehension' is the one word that is generally seen as anchoring Whitehead's theory of perception, its omission is curious.

He had coined the term two years earlier for his Lowell lectures (which would later be published as *Science and the Modern World*) in order to communicate an idea of perception that included non-cognitive entities, for he was convinced that all actual entities (or 'events') *just were* the creative sum of their perceptions,[3] or 'prehensive unification/graded envisagement':

> The word *perceive* is, in our common usage, shot through and through with the notion of cognitive apprehension. So is the word *apprehension*, even with the adjective *cognitive* omitted. I will use the word *prehension* for *uncognitive apprehension*: by this I mean *apprehension* which may or may not be cognitive . . . The things which are grasped into a realised unity, here and now, are not the castle, the cloud, and the planet simply in themselves; but they are the castle, the cloud, and the planet from the standpoint, in space and time, of the prehensive unification. (SMW 69–70)

However, his point made, Whitehead then self-consciously drops the term:

> Now that we have cleared space and time from the taint of simple location, we may partially abandon the awkward term prehension. This term was introduced to signify the essential unity of an event, namely, the event as one entity, and not as a mere assemblage of parts or of ingredients. It is necessary to understand that space-time is nothing else than a system of pulling together of assemblages into unities. But the word *event* just means one of these spatio-temporal unities. Accordingly, it may be used instead of the term 'prehension' as meaning the thing prehended. (SMW 72)

Despite not using the term at all in his Barbour-Page lectures, Whitehead had apparently decided a year later that it was not so awkward after all, or at least that it was less awkward than any alternatives he could imagine, for one can scarcely turn a page of *Process and Reality* without seeing the word 'prehension'.

Symbolism and Error

In *Symbolism*, however, Whitehead instead speaks of two perceptive modes: causal efficacy (a deep non-conscious sense of causal relatedness, common to all entities) and presentational immediacy (pure sense-perception). Importantly, he regards both of these modes as infallible, for without 'the doctrine of a direct experience of an external world', there is only 'solipsism of the present moment' (S 29, 33). Given this 'doctrine of direct experience', Whitehead's task became explaining *how we err* (for Whitehead, it is only in the *combination* of the two perceptive modes by conscious creatures as symbolic reference that error arises). Not only is Whitehead seemingly less concerned with error than most Western philosophers – who are more preoccupied with *truth* – he actually regards error as a positive phenomenon, a source of creativity and novelty.

Higher-grade organisms, far more than lower-grade organisms, contain in their understanding of actuality a reference to ideality (SMW 158–9). They have a goal for themselves, a yearning which manifests in an actual entity's grading of relevance of value of other actual entities and possibilities as factors in its own achievement (SMW 162). Those factors which help to lead towards the organism's ideal of achievement are embraced, while other factors are refused and ignored. Consciousness imagines a novel ideal which becomes the organism's purpose and guides it toward the ideal's realisation in fact (FR 20).

This graded envisagement is how the actual includes what (in one sense) is not-being as a positive factor in its own achievement. It is the source of error, of truth, of art, of ethics and of religion. By it, fact is confronted with alternatives (SMW 176–7).

Of course, the errors of symbolism are usually innocuous as far as survival is concerned. The human race would not survive very long if we could not consistently distinguish an apple from a hand grenade; high-grade organisms can be successful only within a certain margin of error in symbolic reference (S 6). But for Whitehead, creativity springs from the errors of symbolism.

Cosmopolity and Ecology

In a surprising turn of thought – rare in Whitehead's work – *Symbolism* also ventures into the connections of Whitehead's view on events and perception with a problem that his philosophy of organism should help to highlight as a matter of urgent concern today: how to create a society that can continue to socially and ecologically advance in its symbolic perception of reality, allowing for creative reimaginings of symbolism that lead to a deeper understanding and a more responsible shaping of the world. The third part of *Symbolism* not only applies Whitehead's understanding of perception and symbolisation to the organisation of society and the root causes of its breakdown or survival, but also demonstrates with the analysis of three social revolutions – the American, the English and the French – the importance of recognising the deeper layers of social forces inherent in perception and symbolisation. And in a truly grand widening of the horizon, Whitehead expresses the ecological concerns of today by relating them to the ability to create symbolisms that perceive the connective flow of words, and ends with warnings against the human inability to perceive the signs of numbness to the organic embeddedness of language in its layered environments, and to creatively transport this sensibility through ever new and more sophisticated symbolisations.

Rethinking Symbolism

In the decades since Whitehead delivered his lectures, work on and interest in symbols has continued unabated. In Whitehead's own time, Ernst Cassirer produced the highly influential *Philosophy of Symbolic Forms* (1923–9), and this work has of late been the subject of increasing interest among scholars. More recently, the work of Jacques Lacan has left an indelible imprint on Continental thought, and has shaped much of contemporary Continental discourse regarding symbols. From the context of cultural anthropology, Mary Douglas, in her *Natural Symbols* (1970), showed how Émile Durkheim's thought could be used to understand the role of symbols in group formation, and in the years since its publication that book has played a large role in the development of cultural theory. Terrence Deacon, finally, has drawn from contemporary findings of science (brain scans, etc.) to make the case that the ability to use and grasp symbols as symbols is precisely what differentiates humans from non-humans. The problems that were the focus of Whitehead's Barbour-Page lectures are thus very much an ongoing concern.

It was in the context of a lack of scholarship on Whitehead's little book, a resurgence of interest in his philosophy generally, and the need to re-examine our relationship to symbolism in a world that is undergoing massive scientific, cultural and technological changes that the Whitehead Research Project convened its seventh International Whitehead Conference in Claremont, California, in December 2014. It gathered a group of leading scholars from a diverse array of fields in order to examine the current attitude towards symbolism in the modern world. The question of symbolisation and the complex interferences and inferences it establishes in its relationship to modes of perception leads to a discussion of language – whether written or spoken – and to Whitehead's theory of prehension as a way of explaining the construction of human society and (indirectly) ecology. By drawing from a number of disciplines, the conference aimed to set the stage for a fruitful series of discussions on the contemporary meaning and effects of symbolism. The essays now published in this collection illustrate the wide range of ways in which one can engage with Whitehead's thought.

Perception and Paradox

Part I of the present volume focuses on critically examining *Symbolism* while unpacking its persistent puzzles and suggesting a diverse avenue of approaches. The first chapter is by Steven Shaviro, who introduces Whitehead's concepts of perception, causality, symbolism and error with his customary precision and clarity, and effectively contextualises them within the wider Western philosophical canon, including the Kantian and Humean traditions, as well as more recent philosophical developments, such as Graham Harman's object-oriented ontology.

Keith Robinson's chapter, 'Originary Symbolism', undertakes a comparison of Whitehead's views on perception with those of post-structuralist philosopher Gilles Deleuze, who was a great admirer of Whitehead, and who once wrote that 'he stands provisionally as the last great Anglo-American philosopher before Wittgenstein's disciples spread their misty confusion, sufficiency, and terror'.[4] Contrasting Whitehead's account of originary symbolism with Deleuze allows a drawing out of some of the radical innovations and variations of the process view with regard to perception and life.

Finally, in a fitting last chapter for Part I, Roland Faber provides a series of what he calls 'strange attractors' found in *Symbolism*. In Faber's words, 'A "strange attractor" is not strange because that which it reveals in a series of circumambulations is foreign to, or outside of, any expectation, but because it is somehow surprising in its connectivity, a novelty without apparent system of integration'. In approaching Whitehead's book as a series of strange attractors, Faber gestures towards the multitude of approaches that can be fruitfully undertaken in engaging with Whitehead's thought, both teasing and exciting us with the many intriguing loose threads of the tapestry of Whitehead's philosophy.

Adventures in Culture and Value

Part II connects and integrates Whitehead's theory of symbolism with contemporary discussions of culture, politics and value. Michael Halewood's chapter 'The Inhumanity of Symbolism' discusses the ways in which symbolism separates us from the world, which he relates to Marx's concept of the fetishism of the commodity, in which we 'fail to see the human (or social) relations that have gone into making them'. He attacks the problem through a comparison of Whitehead and Marxist Raymond Williams, concluding that the concept of 'ideology' is ultimately a distraction in this discussion, although some degree of 'inhumanity' must always remain inherent in symbolism.

Speaking of inhumanity, Beatrice Marovich's chapter, 'Reverence, Revision and Creaturely Life' explores the role of the 'creature' in Whitehead's philosophy, a word that he applies equally to electrons, humans and God, and primarily serves as a cognate for his 'actual entities'. In analysing Whitehead's use of the word/concept 'creature', Marovich explores his reverence for both the literary tradition of British romantics and the Christian theological tradition, and the way in which this reference leads into a discussion of the creature as value, suggestive as it is of the 'satisfaction' and 'enjoyment' of Whitehead's entities.

Hyo-Dong Lee's chapter, '*Ren* and Causal Efficacy', leaves the Western tradition behind in undertaking a comparison of the social role of symbolism in Whitehead's philosophy with that in Confucianism. For Confucians, rituals symbolically communicate and enhance their emotional responses to their inherently relational lives. With both Whitehead and Confucians stressing a constitutively relational ontology, the Confucian theory of rituals provides an alternative avenue of approach for 'appreciat[ing] Whitehead's implied critique of the modern liberal social theories that are based on a view of human beings as atomised individuals who rationally consent to enter society'.

Sheri Kling then makes an intriguing connection of Whitehead with Carl Jung and dream symbols, citing Whitehead's statement that should we find 'instances of non-sensuous perception, then the tacit identification of perception with sense-perception must be a fatal error barring the advance of systematic metaphysics' (AI 180), and naming dream symbols as just such instances on 'non-sensuous perception'. With this connection, she explores resonances between Whitehead and Jung – including a dipolar God and a collective unconscious – while arguing that an integration of the two can 'positively influence human society's intensity of experience and our overall aliveness, vitality and zest for life'.

Misplaced Concreteness in Ethics and Science

Part III relates Whitehead's theories of symbolism to current controversies in ethics and science, with special attention to the current state of scientific practice, as well as urgent ecological concerns. In the opening chapter, 'A Dog's Life', Jeffrey Bell re-examines *Symbolism* in the light of recent work in the philosophy of language – especially the work of Elisabeth Camp – that has emphasised the importance of 'stimulus-independent' representational abilities in understanding the nature of concepts and the extent to which they play a role in the thoughts of non-humans. He argues that 'deterritorialisation' and 're-territorialisation' are key concepts in understanding thought and the interface between 'the panvitalist processes of life and the panpsychic processes of contemplation'.

Luke Higgins' chapter, 'From Manipulation to Co-Creation', examines the ethically problematic nature of symbol creation, which seemingly forces us either to become subservient to the transcendental symbolism of our religion or culture, or to use symbols to manipulate people and objects in the 'crucible of predatory market capitalism'. He proposes a third way of approaching symbol creation, one which avoids becoming transfixed by the evil of the 'perpetual perishing' of things, and the 'mutually obstructive character of the universe', so that we can take up the challenge of being symbol-makers.

Adam Nocek's chapter, 'On Symbols, Propositions and Idiocies', explores the importance of the figure of the idiot as discussed by Deleuze and Isabelle Stengers, the person who 'protests "what everybody knows" and what passes for "common sense"'. He argues with Stengers for a 'slow science', citing Whitehead's loud protestations about systems of education which teach inert facts within narrow, established fields, education which produces 'minds in a groove', with all the drawbacks that phrase suggests. Science, he argues, has committed Whitehead's 'fallacy of misplaced concreteness' in regarding its symbols as facts, and must reclaim its 'idiotic' passionate objectors.

The final chapter is Catherine Keller's 'Climate Concreteness, Causal Efficacy and the Whiteheadian Cosmopolis'. It asks whether

Whitehead's *Symbolism* can help us to rethink strategies for public education about global warming. In conversation with William Connolly, she marvels at the way our changing weather is now interpreted through a political lens ('Remember', Keller asks, 'when talk about the weather was the most innocuous kind of conversation?') and examines the resources in Whitehead's work which might help to overcome perceptions coloured by political ideologies, including Whitehead's concepts of creativity (his 'category of the ultimate'), beauty as a universal teleology, and radical relationism that 'never washes out difference but intensifies it'.

Final Thoughts – Symbolism and the Evolution of Culture

There is a peculiar dialectical relationship between humans and culture; each of these helps to create the other (RM 87). Each person is a member of the community, and the community is just the composition of its persons. But 'each unit has in its nature a reference to every other member of the community, so that each unit is a microcosm representing in itself the entire all-inclusive universe' (RM 91). Yet the errors inherent in symbolism are fully apparent in the diverse and idiosyncratic ways in which each individual interprets cultural symbols held in common, with unintentional misinterpretations of others' views contributing as much as deliberate modifications and additions. And all people in turn pass along their own unique modifications of the envisagement of the societal ideal – however slight – to their successors, who thus receive a slightly different baseline in adding their own unique insights to the total general outlook. Thus the 'errors' of interpretation created by symbolic reference multiply one on top of another in order to create the phenomenon of social evolution and progress itself. No system of symbols and metaphors can ever hope to encompass the whole of reality, and yet we can always hope to describe it more aptly. As Whitehead tells us,

> Progress in truth – truth of science and truth of religion – is mainly a progress in the framing of concepts, in discarding artificial abstractions or partial metaphors, and in evolving notions which strike more deeply into the root of reality. (RM 131)

But Whitehead also warns us that this revision, however necessary, is often painful and dangerous for the society that undertakes it: 'It is the first step in sociological wisdom, to recognize that the major advances in civilization are processes which all but wreck the societies in which they occur' (S 88). Given this line from the final page of Whitehead's *Symbolism*, this volume of collected essays will hopefully upset established conventions about the way we use language, the way we view animals, the way we do science, and many other things besides.

Notes

1. In a letter to Victor Lowe (Whitehead's biographer), Leonard Woolf wrote that 'My generation regarded Whitehead with respect as co-author with Russell [of *Principia Mathematica*], but we thought he went much too "religious" in his later books'. The Woolfs knew the Whiteheads personally; the character Evelyn Whitbread in Virginia Woolf's *Mrs Dalloway* is most assuredly a caricature of Whitehead's wife, Evelyn. See Woolf, *Letters of Leonard Woolf*, 540.
2. See the letter Whitehead wrote to Charles Hartshorne, reproduced in Kline, *Alfred North Whitehead*.
3. This view has not historically been a very popular one. Many years later, prominent process philosopher and theologian David Ray Griffin would coin the term 'panexperientialism' to describe Whitehead's position and answer critics who spoke disparagingly of Whitehead as a panpsychist. See Griffin, 'Some thoughts on the discussion'.
4. Deleuze, *The Fold*, 76.

Bibliography

Deleuze, Gilles, *The Fold: Leibniz and the Baroque*, trans. Tom Conley (Minneapolis: University of Minnesota Press, 1993).

Griffin, David Ray, 'Some thoughts on the discussion', in John B. Cobb, Jr, and David Ray Griffin (eds), *Mind in Nature: Essays on the Interface of Science and Philosophy* (Washington, DC: University Press of America, 1977), pp. 97–100.

Kline, George L., *Alfred North Whitehead: Essays on His Philosophy* (Lanham: University Press of America, 1989).

Woolf, Leonard, *Letters of Leonard Woolf*, ed. Frederic Spotts (New York: Harcourt Brace Jovanovich, 1989).

Whitehead, Alfred North, *Adventures of Ideas* (New York: The Free Press, [1933] 1967).Whitehead, Alfred North, *Religion in the Making* (New York: Fordham University Press, [1926] 1996).

Whitehead, Alfred North, *Science and the Modern World* (New York: The Free Press, [1925] 1967).

Whitehead, Alfred North, *Symbolism: Its Meaning and Effect* (New York: Fordham University Press, [1927] 1985).

Whitehead, Alfred North, *The Concept of Nature* (Cambridge: Cambridge University Press, [1920] 1971).
Whitehead, Alfred North, *The Function of Reason* (Boston: Beacon Press, [1929] 1958).

Part I
Perception and Paradox

I

Whitehead on Causality and Perception

STEVEN SHAVIRO

Whitehead discusses symbolism – among other reasons – in order to get a handle on the problem of *error*. This, of course, is something that has preoccupied Western philosophy for a long time. Descartes' *Meditations on First Philosophy* begins with his worries about 'how numerous were the false opinions that in my youth I had taken to be true, and how doubtful were all those that I had subsequently built upon them'.[1] Whitehead's erstwhile collaborator Bertrand Russell similarly opens his volume *The Problems of Philosophy* with the question: 'Is there any knowledge in the world which is so certain that no reasonable man could doubt it?'[2] Modern Western philosophy – from Descartes through to Kant and on to today – generally privileges epistemology over ontology. We cannot claim to know the way things are without first giving an account of *how* it is that we know. We cannot consider the consequences of a proposition until we have first assured ourselves that it is free from error.

Whitehead gives his own deceptively bland statement of the problem of truth and error towards the beginning of *Symbolism*: 'An adequate account of human mentality requires an explanation of (i) how we can know truly, (ii) how we can err, and (iii) how we can critically distinguish truth from error' (S 7). Despite this unexceptionable goal, however, Whitehead does not seem to think that the problem of error is of great importance. Indeed, he takes what most philosophers would consider a cavalier, indeed irresponsible, attitude towards the whole question. For he holds that 'in the real world it is more important that a proposition be interesting than that it be true' (PR 259). A scientific observation, a common-sense hypothesis, or even a rigorous philosophical formulation may have relevant and important consequences despite the fact that it is erroneous. For this reason, Whitehead is less concerned with eliminating error than with experimenting with it, and seeing what might arise from it. Error is not an evil to be exterminated, but a frequently useful 'lure for feeling' (PR 25 and passim). It is a productive detour in the pathways of mental life: 'We must not, however, judge too severely of error. In the initial stages of mental progress, error in symbolic reference is the discipline which promotes imaginative freedom' (S 19).

It is worth underlining how rare this position is in Western philosophy. It may well be a cliché of educational method (a subject in which Whitehead himself was deeply interested) that making mistakes is a necessary part of learning. But most philosophers overlook this. They are more concerned with the nature and content of truth than they are with the question of how we may learn to attain it. Deleuze is the only other major philosopher I know who joins Whitehead in regarding the problem of error as in itself merely trivial.[3]

Western philosophy in general is preoccupied with the question of error because it is deeply concerned with the unreliability of immediate experience – or of the body and the senses. From Plato's allegory of the cave, through Descartes' radical doubt about the evidence provided by his physical organs, right on up to Thomas Metzinger's claim that experience is nothing but an internal, virtual-reality simulation, philosophers have been haunted by the idea that sense-perception is delusional – and that, as a result, our beliefs about the world might well be radically *wrong*.

Even if we trust the evidence of our senses, however, we may still be severely limited in the extent of what we can actually know. Hume is sceptical, not so much of the deliverances of the senses themselves, as of what we can legitimately infer from them. For Hume, 'all events seem entirely loose and separate. One event follows another; but we never can observe any tie between them. They seem *conjoined* but never *connected*.'[4] It is true that we often observe the 'constant conjunction' of certain events. But correlation is not causation, and we cannot legitimately infer from the former to the latter. Hume concludes that the 'idea of a necessary connexion among events' arises only because 'the mind is carried by habit' to expect a second, associated event when it encounters the first.[5]

Kant, of course, endeavours to overcome Hume's scepticism by means of a transcendental argument. We cannot do without causality. If relations of cause and effect cannot be found in sense-data themselves, as Hume maintains, then they must inhere in 'our ways of thought about the data' (S 37). For Kant, causality is rescued as an *a priori* category of understanding. If we were not able to organise the sense-data we receive according to the laws of cause and effect, Kant says, then we would scarcely be able to have subjective experience at all.

Recent philosophy most often treats causality in a Humean spirit, rather than a Kantian one. Thus the late analytic philosopher David Lewis maintains that 'all there is to the world is a vast mosaic of local matters of particular fact, just one little thing and then another'.[6] Relations of cause and effect may be observed to *supervene* upon these particular facts; but Lewis argues, following Hume, that we cannot make any inference from such observations to a deeper sort of necessity. For we can always imagine, without logical contradiction, counterfactual *possible worlds* in which events could have turned out differently. Analytic philosophers love to float scenarios in which, for instance, water is not H_2O,[7] or people devoid of sentience nonetheless act in ways that are indistinguishable from everyone else.[8] Indeed, Lewis's 'modal realism' asserts that we must accept the reality of all these alternative possible worlds.

As Jeffrey Bell has noted,[9] there is a certain similarity between Lewis's doctrine of Humean supervenience and the revival, by the speculative realist philosopher Quentin Meillassoux, of what he calls 'Hume's Problem'.[10] For Meillassoux, Hume establishes once and for all that neither experience (which pertains only to the past and present, never to the future) nor *a priori* reasoning (which can only exclude logical contradictions) is able to guarantee the necessity of causal relations. For 'there is nothing contradictory in thinking that the same causes could produce different effects tomorrow'.[11] If the prospect of arbitrary change is not impossible, Meillassoux argues, then it cannot be excluded from the world as it is. Where Lewis affirms the reality of all possible worlds, Meillassoux argues for 'the absolute necessity of contingency', or of sheer ungrounded possibility, in our own world.[12]

Hume and Kant alike, as well as their followers, share what Whitehead calls the 'naïve presupposition of "simple occurrence" for the mere data' – or better, of 'simple location', since it applies 'to space as well as to time' (S 38). It little matters for Whitehead, therefore, whether 'causal efficacy' is defined with Hume as 'a habit of thought' or with Kant as 'a category of thought' (S 39–40). In both cases, relations and forms of organisation are abstracted away from the matrix of things themselves, and attributed only to the mind that observes these things. 'Both schools find "causal efficacy" to be the importation, into the data, of a way of thinking or judging about those data' (S 39).

Whitehead, however, rejects the presuppositions that underlie this whole history of argument. For Whitehead denies that events in themselves are ever merely 'loose and separate', or that the world can be reduced to 'local matters of particular fact'. In the actual world, he says, 'there is nothing which "simply happens"' (S 38). There are no isolated data, because in every act of experience 'the datum includes its own interconnections' already (PR 113). In order to explain how this works, Whitehead distinguishes between two separate modes of perceptive experience: *presentational immediacy* and *causal efficacy*. These two modes, together with the ways that they are fused in symbolic reference, form the main subject of *Symbolism*. The distinction between these two modes is further elaborated in *Process and Reality*.

Presentational immediacy roughly corresponds to Descartes' 'clear and distinct perceptions', Hume's 'impressions' and Kant's 'sensible intuitions'. Whitehead defines it as 'our immediate perception of the contemporary external world', an appearance 'effected by the mediation of qualities, such as colours, sounds, tastes, etc.' (S 21). Presentational immediacy is the great source of sensuous richness. But it provides us only with clearly demarcated representations, and it is confined to the present moment, without any thickness of duration. For these reasons, presentational immediacy is severely limited in what it reveals of the world. As Whitehead says, presentational immediacy is 'vivid, precise, and barren' (S 23). It 'displays a world concealed under an adventitious show, a show of our own bodily production' (S 44). But for this very reason, it leaves us with a hollow sense of depthless mere appearances. This is the root of philosophical scepticism, in Hume and throughout modernity.

According to Whitehead, the problem with standard philosophical accounts of perception is that these accounts are concerned *only* with presentational immediacy. They entirely ignore other modes of experience. They take it for granted that our empirical experience is limited to individual sense impressions, or to the 'local matters of particular fact' that correspond to these impressions. This assumption is what allows Hume to argue that objects are nothing more than hypothetical bundles of qualities. It is also what drives Kant to conclude that only the mind can bring order to what would otherwise be a chaos of unrelated impressions.

Whitehead, however, suggests that Hume and Kant do not even give presentational immediacy its proper due. For he insists that, even if we restrict ourselves to just this mode of perception, 'the world discloses itself to be a community of actual things, which are actual in the same sense as we are' (S 21). When we are looking at a wall, for instance, 'our perception is not confined to universal characters; we do not perceive disembodied colour or disembodied extensiveness: we perceive *the wall*'s colour and extensiveness' (S 15, original emphasis). Contrary to the empiricist assumption of separate, atomistic qualia, in fact 'there are no bare sensations which are first experienced and then "projected" into our feet as their feelings, or onto the opposite wall as its colour' (S 14). The supposedly atomistic, qualitative sense-data are not initially isolated from one another. Rather, Whitehead says, such qualities 'can be thus isolated only by abstracting them from their implication in the scheme of spatial relatedness of the perceived things to each other and to the perceiving subject . . . the sense-data are generic abstractions' (S 22).

It is worth noting that Graham Harman, with his object-oriented ontology, also opposes what he describes as 'the widespread empiricist view that the supposed objects of experience are nothing but bundles of qualities'. Harman rather insists that qualities are never isolable, but always 'bonded to the thing to which they belong'.[13] Harman attributes this point to Husserl, for whom an 'intentional object' is not the sum of its adumbrations, but always more than its multiple aspects or qualities.[14] 'According to Husserl we encounter the intentional object directly in experience from the start'; it does not have to be 'built up as a bundle of perceptually discrete shapes and colors, or even from tiny pixels of sense experience woven together by habit'.[15]

My reason for mentioning this is that Whitehead makes the same distinction as Husserl does – at least according to Harman's reading of Husserl. Whitehead most likely makes this point without having encountered it in Husserl. It is true that Whitehead had students – most notably Charles Hartshorne – who had also studied with Husserl and were familiar with his writings. But I do not see any evidence for Husserl's influence upon Whitehead, even when – as here – they come to parallel conclusions.

Be that as it may, for Whitehead the major defect in mainstream philosophical accounts of perception is that they leave out any consideration of causal efficacy. The physical sciences, on the other hand, are predominantly concerned with causal efficacy, but they treat it only as an objectified process, comprehended by a 'view from nowhere'. In this way, the split between presentational immediacy and causal efficacy is a prime instance of what Whitehead calls the *bifurcation of nature* (CN 26–48). The scientists, no less than the philosophers, neglect causal efficacy as a form of perception, or as a mode of experience. It is only by treating causal efficacy experientially, and understanding how it becomes entwined with presentational immediacy in the operations of symbolic reference that we can overcome the opposition between phenomenology and natural science, or between 'the nature apprehended in awareness and the nature which is the cause of awareness' (CN 31).

Whitehead goes to great lengths in *Symbolism* to argue not only that causal efficacy is a mode of perception, but also that it is the most primordial mode of perception, far deeper than presentational immediacy. The latter 'is only of importance in high-grade organisms' (S 16). But 'the direct perception of causal efficacy' operates everywhere (S 39). For it involves 'the overwhelming conformation of fact, in present action, to antecedent settled fact' (S 41). Indeed, Whitehead says, 'the perception of conformation to realities in the environment is the primitive element in our external experience. We conform to our bodily organs and to the vague world which lies beyond them' (S 43).

Without this conformation of the present to the past, this physical experience of causal efficacy, the clarities and intensities of presentational immediacy could not even arise for us in the first place. Even our most clear and distinct perceptions are grounded in a deeper sense that is 'vague, haunting, unmanageable' (S 43). Our very awareness of sharp and delicious sensations, and our ability to make subtle discriminations among them – what Whitehead describes as our 'self-enjoyment derived from the immediacy of the show of things' – are underwritten and made possible by 'the perception of the pressure from a world of things with characters in their own right, characters mysteriously moulding our own natures' (S 44). A heavy otherness insinuates itself into even our clearest and most distinct perceptions, which is why there can be no 'solipsism of the present moment' (S 29).

This massive underlying pressure of causal efficacy is also what produces and accounts for our apprehension of things as more than just bundles of qualities:

> these primitive emotions are accompanied by the clearest recognition of other actual things reacting upon ourselves. The vulgar obviousness of such recognition is equal to the vulgar obviousness produced by the functioning of any one of our five senses. When we hate, it is a man that we hate and not a collection of sense-data – a causal, efficacious man. (S 45)

The vagueness of the emotional experience of causal efficacy does not prevent, but rather actually calls forth, an awareness that things actually do exist outside us and apart from us. In other words, 'we encounter the . . . object directly in experience from the start', as Harman insists, rather than building up a representation of the object from a bundle of separate sense impressions. My direct experience of the object in the mode of causal efficacy subtends my identification of it in the mode of presentational immediacy. And it is only by abstracting away from causal efficacy, with its 'overwhelming conformation of fact, in present action, to antecedent settled fact' (S 41), that we can enjoy the subtle and disinterested aesthetic pleasures of presentational immediacy.

This is why, following Whitehead, I dissent from Harman's insistence that 'real objects cannot touch',[16] and that causation can only be 'vicarious'.[17] For this is the case only from the viewpoint of presentational immediacy. In causal efficacy, objects *do* literally touch one another. This immediacy of touch follows directly from 'the principle of conformation, whereby what is already made becomes a determinant of what is in the making . . . The present fact is luminously the outcome from its predecessors, one quarter of a second ago' (S 46). The principle of conformation applies equally to my own continuity with whom I was a quarter of a second ago, and to my contact with things that have impinged upon me in the past quarter second.

Harman worries that all distinction would be lost if actual contact were possible. He argues that the idea 'of indirect-but-partial contact cannot work . . . *Direct* contact could only be all or nothing'.[18] Harman's problem is to maintain separation at the same time that he accounts for causal influence. As Harman puts it, even when fire burns cotton, there is no direct contact between these two entities. The fire may well obliterate the cotton, with no remainder. But even then, Harman says, 'fire does not interact at all' with such qualities as 'the cotton's odor or color'.[19] Therefore fire and cotton remain ontologically separate, in accordance with Harman's dictum that 'the object is a dark crystal veiled in a private vacuum'.[20]

Now, Isabelle Stengers insists that Whitehead always works as a mathematician, even when he is engaged in philosophical speculation. Whitehead does not posit absolute principles; rather, he always confronts specific problems, by producing a construction that observes all 'the constraints that the solution will have to satisfy'.[21] In this sense, Whitehead's distinction between presentational immediacy and causal efficacy is itself constructed as a way to resolve the problem of error and the scepticism about causality that are found in the Humean and Kantian traditions.

I would like to suggest that, in this way, Whitehead offers a construction that resolves what I have just called Harman's problem. He argues that, at one and the same time, 'actual things are *objectively* in our experience and *formally* existing in their own completeness . . . no actual thing is "objectified" in its "formal" completeness' (S 25–6). This allows him to assert both:

1. that things actually do enter into direct contact with other things, as they partially determine the composition of those other things; and
2. that no particular thing is entirely subsumed, either by the other things that entered into it and helped to determine its own composition, nor by the other things into which it subsequently enters.

In this way, Whitehead's construction satisfies – ahead of time – all the conditions of Harman's problem, without accepting Harman's vision of objects as inviolable substances. I will note as well that Whitehead's reappropriation of the old scholastic distinction between 'formal' and 'objective' existence has an affinity with Tristan Garcia's version of object-oriented philosophy, according to which a thing is defined as the difference between '*that which is in a thing* and *that in which a thing is*, or that which it comprehends and that which comprehends it'.[22] Garcia, like Whitehead, refuses to explain away causal efficacy, while at the same time recognising what Whitehead calls 'the vast causal independence of contemporary occasions' which 'is the preservative of elbow-room within the Universe. It provides each actuality with a welcome environment for irresponsibility' (AI 195).

The larger point here is that causal efficacy is at one and the same time a mode of perception and an actual physical process. It encompasses both 'the perceived redness and warmth of the fire' and 'the agitated molecules of carbon and oxygen . . . the radiant energy from them, and . . . the various functionings of the material body' (CN 32). In this double functioning, causal efficacy is irreducible to rigid determinism, but also impregnable to philosophical scepticism.

Whitehead notes, for instance, that Hume's own presuppositions contradict his assertion that causal efficacy cannot be directly perceived:

> Hume with the clarity of genius states the fundamental point, that sense-data functioning in an act of experience demonstrate that they

are given *by* the causal efficacy of actual bodily organs. He refers to this causal efficacy as a component in direct perception. (S 51, original emphasis)

That is to say, by Hume's own prior admission we get direct acquaintance with the world through the actions of the body.

> In asserting the lack of perception of causality, [Hume] implicitly presupposes it... His argument presupposes that sense-data, functioning in presentational immediacy, are 'given' by reason of 'eyes', 'ears', 'palates' functioning in causal efficacy. (S 51)

More generally, Whitehead says,

> we see the picture, and we see it with our eyes; we touch the wood, and we touch it with our hands; we smell the rose, and we smell it with our nose; we hear the bell, and we hear it with our ears; we taste the sugar, and we taste it with our palate. (S 50)

The functioning here of experience in the mode of causal efficacy is antecedent to, and necessary for, the very experience in the mode of presentational immediacy within which, Hume says, no causation can be discerned.

Whitehead recapitulates and expands this critique of Hume in *Process and Reality*. Hume argues that our expectation that a certain effect will follow a cause is merely a product of habit. But Whitehead notes that

> it is difficult to understand why Hume exempts 'habit' from the same criticism as that applied to the notion of 'cause'. We have no 'impression' of 'habit', just as we have no 'impression' of 'cause'. Cause, repetition, habit are all in the same boat. (PR 140)

Once again, Hume presupposes the power of causal efficacy in his very attempt to explain it away.

I am tempted to describe Whitehead's mode of argument here as a precise inversion of Kant's. Kant opposes Hume by insisting that we cannot, in principle, escape causality, because it must be imposed transcendentally from above. Whitehead instead opposes Hume by observing that, in point of fact, we do not escape causality because it is always already at work empirically, from below. Whitehead turns Kant around and puts him on his feet, we might say, in the same way that Marx put Hegel on his feet.

Whitehead shows that causal efficacy is always already at work in our perception, as a physical functioning of the bodily organs. This would remain the case even if we were brains in vats, getting delusive sense impressions by means of direct stimulation of the neurons. The actual physical functioning of causal efficacy must still be presupposed, even if the picture presented through presentational immediacy does not correspond to an actual state of affairs in the world.

This is why Whitehead says that 'direct experience' in itself 'is infallible'. This assertion is in fact a tautology: 'what you have experienced, you have experienced' (S 6). The delusion of a brain in vats, like the delusion exhibited in 'Aesop's fable of the dog who dropped a piece of meat to grasp at its reflection in the water' (S 19), is a failure of symbolic reference, rather than of direct experience in itself. It results not from any defect of perception per se but from the way in which 'the various actualities disclosed respectively by the two modes are either identified, or are at least correlated together as interrelated elements in our environment' (S 18).

In other words, the dog's error is a mistake of interpretation, or a failure to respect the limits of abstraction. Whitehead tells us that we cannot live without making abstractions, even though we go wrong when we take our abstractions too seriously, or push them beyond the limits within which they are useful. This is what Whitehead famously calls 'the fallacy of misplaced concreteness' (S 39); we find it at work not just in a dog's misjudgement, but also in the most refined examples of philosophical reasoning. It is not the perception of meat in the water that is at fault, but rather the dog's failure to understand that this meat – which he truly perceived – is a reflection rather than an edible substance. This is why Whitehead remains so relaxed in his treatment of error: 'Aesop's dog lost his meat, but he gained a step on the road towards a free imagination' (S 19).

We experience causal efficacy not only because we are bodies, but also because we feel, and subsist within, the passage of time. Whitehead argues that Hume's sceptical conclusions 'rest upon an extraordinary naïve assumption of time as pure succession' (S 34). This notion of 'pure succession', or time as an empty form, 'is an abstraction from the irreversible relationship of settled past to derivative present' (S 35). In actual concrete experience, we feel time as 'the derivation of state from state, with the later state exhibiting conformity to the antecedent . . . The past consists of the community of settled acts which, through their objectifications in the present act, establish the conditions to which that act must conform' (S 35).

In other words, experience does not only happen in the present moment, in the Now. It also comprehends the past, and projects towards the future. Even the most 'primitive living organisms . . . have a sense for the fate from which they have emerged, and for the fate towards which they go' (S 44). Time is not so much the measure of change as it is the force of 'conformation'; and it is only against the background of this force of conformation that change is even possible:

> the present fact is luminously the outcome from its predecessors, one quarter of a second ago. Unsuspected factors may have intervened; dynamite may have exploded. But, however that may be, the present event issues subject to the limitations laid upon it by the actual nature of the immediate past. If dynamite explodes, then present fact is that issue from the past which is consistent with dynamite exploding. (S 46)

In this way, perception and judgement are themselves temporal instances. They are nested within the broad span of 'conformation' or causal influence. To perceive something is to be affected or influenced by that something. And willed action – or more generally, what Whitehead in *Process and Reality* calls *decision* (PR 27–8 and passim) – can itself take place only within a given framework of causal efficacy. This is the source of Whitehead's distinction, in *Symbolism*, between 'pure potentiality' and 'natural potentiality' (S 37–8) – which is recast in *Process and Reality* as a distinction between 'general potentiality' and 'real potentiality' (PR 65). Pure or general potentiality is mere logical possibility, while natural or real potentiality takes account of 'stubborn fact', or of the actual 'components which are *given* for experience' (S 36, original emphasis).

From a Whiteheadian point of view, Lewis's modal realism and Meillassoux's principle of contingency both fail because they ignore this distinction. Since they recognise only presentational immediacy, they abstract 'the mere lapse of time' from 'the more concrete relatedness of "conformation"' (S 36). In consequence, they regard sheer logical possibility – what Whitehead calls pure or general potentiality – as if it were natural or real potentiality. 'According to Hume', Whitehead says, 'there are no stubborn facts' (S 37), and the same must be said for Lewis and Meillassoux. The error of these great thinkers, we might say, results precisely from their endeavour to eliminate error on grounds of epistemological consistency.

For the mainstream of modern Western philosophy, causality is an example of a relation that must be put into doubt, because it is supposedly not given in perception. Whitehead counters this by showing that causality is not just an abstract condition for perceptive experience (which Kant had argued already), but is also an actually given component of experience. Causal efficacy is in fact directly experienced, even though this direct experience is not necessarily conscious. In *Process and Reality*, Whitehead gives this in the form of an example:

> [When] occasions A, B, and C enter into the experience of occasion M, [this means that] there is thus a transmission of sensation emotion from A, B, and C to M. If M had the wit of self-analysis, M would know that it felt its own sensa, by reason of a transfer from A, B, and C to itself. Thus the (unconscious) direct perception of A, B, and C is merely the causal efficacy of A, B, and C as elements in the constitution of M. (PR 115–16)

Causal efficacy is itself experienced in a vague and limited way: it is thus a primordial form of perception. But beyond this, experience of any sort materially depends upon the functioning of causal efficacy. Therefore, causality is more than just an *example* of something whose status in perception we may argue about. In fact, it is central to the whole theory of perception. Perception is itself a sort of causal relation – rather than causal relations being instances that we may perceive or not.

In this way, Whitehead's account of causal efficacy provides a bridge from epistemology to ontology, or to what Whitehead calls cosmology. For Hume, Kant and their modern successors, we cannot talk about causality without first accounting for *how we know* that causal relations between ostensibly independent entities can exist. But Whitehead argues that even to raise the question of *how we know* is already to have accepted the operation of causal efficacy, in the form of the 'conformation of present fact to immediate past' (S 41). Whitehead thus cuts the Gordian knot of Kantian critique; he frees speculation from the grim Kantian alternative of either

1. being subjected to critique, which is to say to prior epistemological legitimation, or
2. being rejected as simply 'dogmatic'.

It should be noted that Quentin Meillassoux also seeks to escape this infernal alternative. He claims to establish the possibility of 'non-dogmatic speculation'[23] as a way of stepping outside the Kantian 'correlationist circle'[24] without thereby performing a '*pre-critical* . . . regression to the "naïve" stance of dogmatic metaphysics'.[25] Whitehead describes his own speculative philosophy as 'a recurrence to that phase of philosophic thought which began with Descartes and ended with Hume' (PR xi). Nonetheless, I do not think that Whitehead's constructivist proposal for solving the riddles of perception and causality can be categorised as 'dogmatic' in the pejorative Kantian sense. Rather, Whitehead's speculative 'flight in the thin air of imaginative generalization', together with his subsequent return to the ground 'for renewed observation rendered acute by rational interpretation' (PR 5), allows him to perform what he describes, in another act of setting Kant on his feet, as 'the self-correction by consciousness of its own initial excess of subjectivity' (PR 15). This is why I have sought to establish a dialogue between Whitehead, on the one hand, and recent speculative realist thinkers like Meillassoux and Harman, on the other. It seems to me that Whitehead anticipates many of the goals of the speculative realists. At the same time, Whitehead offers an alternative both to Meillassoux's excessive rationalism and to Harman's grounding in phenomenology.

I will conclude by mentioning some further consequences of this discussion, even though I cannot fully explore them here. Whitehead argues both that causal efficacy is directly perceived and that the causal conformation of the present to the immediate past is a general process, of which direct perception in either mode is just an example. There is therefore a curious chiasmus between perception and causality, which intersect in something like a feedback loop. This also implies, among other things, that there is no clear dividing line between perception proper and causal influence more generally. I 'perceive' something whenever I am affected by that something – even in cases where this does not happen consciously. For instance, Whitehead notes that 'the human body is causally affected by the ultra-violet rays of the solar spectrum in ways which do not issue in any sensation of colour. Nevertheless such rays produce a decided emotional effect' (S 85).

This 'emotional effect' may well be a modulation of my mood: I always feel better when I am outdoors on a sunny day. But it may also consist in my getting sun tanned, or sunburnt, or even developing skin cancer. Any physical response of this sort is in some sense an 'emotional' response as well. Even below the threshold of consciousness, a physical change is also a change of some sort in affective tone. This is not only the case for human experience, but also for organisms that Whitehead calls 'low grade', as when 'a flower turns to the light', or even when 'a stone conforms to the conditions set by its external environment' (S 42).

A lot of this has been covered in recent writings on Whitehead under the rubric of what he calls, in *Adventures of Ideas*, 'non-sensuous perception' (AI 180ff.). 'In human experience', Whitehead writes, 'the most compelling example of non-sensuous perception is our knowledge of our own immediate past' (AI 181). All this is consistent with what Whitehead says in *Symbolism* about perception in the mode of causal efficacy. But Mark B. N. Hansen, in his book *Feed-Forward*, argues that such an understanding of Whitehead's expanded field of perception sells him short. Hansen urges us to consider the causal efficacy of 'nonperceptual sensibility' beyond the confines of personal memory, referring to the ways in which causal efficacy extends 'beyond perception' to a domain that '*does not and cannot appear through (human perception)*', but that human beings are now for the first time able to access '*indirectly* . . . through the technical supplement afforded by biometric and environmental computational sensing'.[26] Whitehead's expanded theory of perception is thus crucial, Hansen says, for grasping our emerging twenty-first-century media environment. I have serious disagreements with Hansen's particular interpretation of Whitehead, but I think his overall point is enormously important, and it can be grasped in the terms that I am working through here: the chiasmic relation between perception and physical causality.

On my reading of Whitehead, perception is a subset of causal processes more generally, while at the same time causal processes are themselves 'felt', even unconsciously, as they are fed back into direct perceptual experience. This is the basis for what David Ray Griffin calls Whitehead's *panexperientialism* – though I prefer to use the more provocative word *panpsychism*. This means that differences in mentality, or in levels of what Whitehead calls 'feeling' (using this word as 'a mere technical term') (PR 164), are always differences in degree, rather than in kind. There is no clear boundary line between the different modes of feeling or sentience, just as 'there is no absolute gap between "living" and "non-living" societies' (PR 102).

But I think that we can go further than this. Whitehead says that 'life lurks in the interstices of each living cell, and in the interstices of the brain' (PR 105–6). But feeling – or perception as conformation – does not need to lurk in the interstices; it happens everywhere. This is why I do not think that Whitehead is really a vitalist. Whitehead's conflation of perception with causal efficacy also implies the priority of sentience over vitality. In other words, perception and feeling are among the necessary conditions of possibility for life, rather than life being a necessary condition of possibility for sentience.

Why is this important? As Eugene Thacker has demonstrated at length, in his great book *After Life*, all our attempts to reinvent vitalism, to explore the possibilities of what Deleuze and Guattari call 'inorganic life', and to theorise 'Life' in general, come up against a series of crippling antinomies. In the actual practices of contemporary biotechnology, as well as in philosophical argumentation, Thacker says, 'thought and life approach a horizon of absolute incommensurability; the thought of life becomes increasingly disjunctive with the vague set of phenomena we call "life itself"'.[27] There are contradictions both between particular instances of life and 'life' as an essence or overall concept, and between all these iterations of life and the thought, itself alive, which tries to grasp and conceptualise it. I suspect – though it is only a hunch at this point – that approaching life from the point of view of sentience or feeling, rather than taking sentience as an attribute of life, might help to offer us a way out from these confusions.

Notes

1 Descartes, *Meditations*, 13.
2 Russell, *The Problems of Philosophy*, 6.
3 Deleuze, *Difference and Repetition*, 148–51.
4 Hume, *An Enquiry*, 68.
5 Hume, *An Enquiry*, 69.
6 Lewis, *Philosophical Papers, Volume II*, ix.
7 Putnam, 'Meaning and reference'.
8 Chalmers, *The Conscious Mind*, 93–122.
9 Bell, 'Between realism and anti-realism'.
10 Meillassouox, *After Finitude*, 82–111.
11 Meillassoux, *After Finitude*, 87.
12 Meillassoux, *After Finitude*, 65.
13 Harman, *The Quadruple Object*, 11.
14 Harman, *The Quadruple Object*, 24–5.
15 Harman, *The Quadruple Object*, 25.
16 Harman, *The Quadruple Object*, 73.
17 Harman, *The Quadruple Object*, 128.
18 Harman, *Bells and Whistles*, 34.
19 Harman, *The Quadruple Object*, 44.
20 Harman, *The Quadruple Object*, 47.
21 Stengers, *Thinking with Whitehead*, 33.
22 Garcia, *Form and Object*, 11, original emphasis.
23 Meillassoux, *After Finitude*, 79.
24 Meillassoux, *After Finitude*, 5.

25 Meillassoux, *After Finitude*, 3, original emphasis.
26 Hansen, *Feed-Forward*, 24, original emphasis.
27 Thacker, *After Life*, ix–x.

Bibliography

Bell, Jeffrey, 'Between realism and anti-realism: Deleuze and the Spinozist tradition in philosophy', *Deleuze Studies*, 5:1 (2011), 1–17.

Chalmers, David, *The Conscious Mind: In Search of a Fundamental Theory* (New York: Oxford University Press, 1997).

Deleuze, Gilles, *Difference and Repetition*, trans. Paul Patton (New York: Columbia University Press, 1994).

Descartes, René, *Meditations on First Philosophy*, trans. Donald A. Cress (Indianapolis: Hackett, 1993).

Garcia, Tristan, *Form and Object: A Treatise on Things*, trans. Mark Allan Ohm and Jon Cogburn (Edinburgh: Edinburgh University Press, 2014).

Hansen, Mark B. N., *Feed-Forward: On the Future of Twenty-First-Century Media* (Chicago: University of Chicago Press, 2015).

Harman, Graham, *Bells and Whistles: More Speculative Realism* (Winchester: Zero Books, 2013).

Harman, Graham, *The Quadruple Object* (Winchester: Zero Books, 2011).

Hume, David, *An Enquiry Concerning Human Understanding: And Other Writings* (New York: Cambridge University Press, 2007).

Lewis, David, *Philosophical Papers, Volume II* (New York: Oxford University Press, 1987).

Meillassoux, Quentin, *After Finitude: An Essay on the Necessity of Contingency*, trans. Ray Brassier (New York: Continuum, 2008).

Putnam, Hilary. 'Meaning and reference', *Journal of Philosophy*, 70:8 (1973), 699–711.

Russell, Bertrand, *The Problems of Philosophy* (Radford, VA: Wilder Publications, [1912] 2011).

Stengers, Isabelle, *Thinking with Whitehead: A Free and Wild Creation of Concepts*, trans. Michael Chase (Cambridge, MA: Harvard University Press, 2011).

Thacker, Eugene, *After Life* (Chicago: University of Chicago Press, 2010).

Whitehead, Alfred North, *Adventures of Ideas* (New York: The Free Press, [1933] 1967).

Whitehead, Alfred North, *Process and Reality: An Essay in Cosmology* (New York: The Free Press, [1929] 1978).

Whitehead, Alfred North, *Symbolism: Its Meaning and Effect* (New York: Fordham University Press, [1927] 1958).

Whitehead, Alfred North, *The Concept of Nature* (Amherst: Prometheus Books, [1920] 2004).

2

Originary Symbolism: Whitehead, Deleuze and the Process View on Perception

KEITH ROBINSON

Introduction

Perhaps *the* problem that drives Whitehead's philosophy of nature and his metaphysics is the relation between internal and external standpoints, in particular the relation between the subjective viewpoint 'here' and the objective 'view from nowhere' – what Whitehead famously called 'the bifurcation of nature'. All of Whitehead's metaphysical concepts are constructed with this problem in mind, and his theories of symbolism and perception are no exception. In fact, by tracing out Whitehead's understanding of perception, we follow one route through the problem of bifurcation.

As is well known, Whitehead responds to bifurcation in his later works with his 'one genus' theory of 'dipolar' actual occasions designed to circumvent or escape the difficulties associated with the various dualisms and materialisms of the tradition. Rather than 'panpsychism', which Whitehead never fully subscribed to (at least if we define panpsychism as a generalisation of psyche, mind, or consciousness), Whitehead's theory ascribes a 'physical pole' to every occasion, as well as a more or less recessive 'mental pole'. Thus experience or perception (and, as we will see, Whitehead will generalise and equate them) is a contrast – an integration and synthesis – of physical inheritance and a more or less conceptual reaction. Conceptual appetition here should not be identified with consciousness. For Whitehead, consciousness presupposes experience. Consciousness is contingent and derivative, an evolutionarily later form of integration. Moreover, as a fully fledged 'process theory', the Whiteheadian occasion does not just passively 'have' or 'undergo' experience, is not just a static perceptual experience 'of' the real in the manner of a substance qualified by predicates, but *is itself* experience, an active experience of passage and becoming between interrelated processes that 'influence each other, require each other and lead on to each other' (MT 157). The world is in the occasion and the occasion is in the world. This is Whitehead's Leibnizian-inspired doctrine of 'mutual immanence' (MT 157), the creative 'reciprocal insistence' (PNK 14), as he says, between the occasion and the rest of nature.

Whitehead's accounts of perception are among his most important philosophical legacies because they challenge the bifurcation of nature and attempt to show the connectedness of occasions. Along with Bergson's 'duration' and James's 'stream of consciousness', Whitehead's notions of 'causal efficacy', 'presentational immediacy' and 'symbolic reference' offer a direct challenge to the various schools of thought derived from Hume and Kant in which causation is seen as a pale derivation from the 'sensationalist' vivid impressions of immediate atomic sense-data presented to consciousness. By focusing only on sense-data, we end up in what Whitehead, following Santayana, called the 'solipsism of the present moment' (S 29). In causal efficacy, by contrast, 'the presentations of sense fade away and we are left with vague feelings of influences from vague things around us' (PR 176). These vague influences attest to the repetition of the obscure processes of the past in us, out of which emerge the selective crisp immediacies of our present experience. Whitehead ties his account of perception not only to a certain conception of temporality, but also to a generalised or what I call an *originary* account of symbolism.[1] The logic of Whitehead's originary or 'arche-symbolism' is grounded in the claim that no symbol can be absolutely present and immediate, but is always divided by an internal time that preserves or 'conforms' to the past and opens onto an indeterminate future. Every symbol is conditioned by a temporal moment (or symbolising process) which is always ceasing to be, or what Whitehead calls 'perpetually perishing'. But, in the perishing of the 'now', something is preserved in the past that can be altered or destroyed in the future. There is no 'simple occurrence' and no 'simple location', only the bare or minimal sense in which a symbol always 'references' something else. Originary symbolism is the power to affect or be affected, an exposure to what happens as the condition not just for language, experience, or even God in Whitehead's sense, but for all becoming and life.[2]

This critique of 'natural' perception and the generalisation of an originary differential structure is also taken up and developed in great detail and complexity in the work of the French philosopher Gilles Deleuze. For Deleuze, life is an immense flowing movement of vague and obscure images, singularities and intensities in the midst of things, the prehension of one by the other or the passage and communication from one to the other, such that consciousness is already a becoming immersed in things rather than a being independent of them. Contrasting Whitehead's account of originary symbolism with Deleuze will enable us to draw out some of the radical innovations and variations of the process view with regard to perception, time and becoming, and the implications these have for thinking about immanence without bifurcation.

Whitehead and Symbolism

In his lectures to the University of Virginia, published in 1927 as *Symbolism: Its Meaning and Effect*, Whitehead distinguishes various kinds of symbolism and their prevalence, in particular 'epochs of civilization' (S 2). In some periods, symbolisms of a certain type permeate everything; in other periods, types of symbolism are dispensed with in favour of more direct modes of apprehension. Whitehead briefly mentions language and mathematics in the opening sections and he returns to the example of language at the end. Linguistic and mathematical symbolisms are regarded as 'deeper' than other types of symbolism, such as those used in heraldry or the cathedrals of medieval Europe. Indeed, language and mathematics cannot simply be discarded, since 'we could not get on without them' (S 2). However, Whitehead puts them to one side for now, because he wants to discuss a symbolism even more 'fundamental' (S 3) than language or mathematics.

We look up and see a coloured shape. We say we have seen a 'chair', but we have seen a coloured shape. An artist may not have jumped to the notion of a chair. But if we are tired, we might go from the perception of the coloured shape to enjoying the chair, just as a puppy dog might. We can say that we made an inductive inference, based on past experience of similar shapes, to the conclusion that this is a chair before me. But Whitehead says he is 'sceptical as to the high-grade mentality required to get from the colored shape to the chair' (S 3). He is sceptical because his artist friend was highly trained, and trained to ignore the chair in favour of the coloured shape. We do not need, Whitehead says, lots of training to refrain from complex sequences of inference. On perceiving the coloured shape, the puppy dog may also have jumped on the chair to enjoy it. If the dog did not do this, it might have been because it was trained. The transition from coloured shape to an object that may have many uses seems to be a natural one. Following on from some of the phenomenologists, we might say that the transition that Whitehead refers to here takes place in a 'pre-reflective' or phenomenal field, a 'being in the world' that is embedded within a large frame of references or contexts that allow the coloured shape to disclose itself as a chair, or at least as an object that has certain meanings, values and uses to and for an organism within its world. To disregard those uses and meanings, or to put them to one side, requires 'careful training' (S 4).

The coloured shape emerges from a more primordial perceptive experience that Whitehead calls the 'withness of the body' (PR 64, 81). In this mode of perception 'we see the contemporary chair, but we see it *with* our eyes, and we touch the contemporary chair but we touch it *with* our hands' (PR 62; original emphasis here and below). These statements for Whitehead show direct knowledge, or 'direct recognition' (S 7) as he says in *Symbolism*, of the antecedent functioning of the body and the presence of the world in our experience. Direct knowledge or recognition is not hidden behind appearances. By attending to what appears or what is given in experience, direct knowledge shows itself.

For Whitehead the withness of the body here is not simply the recognition of a material object extended in space. The withness of the body shows a *how I am in the world*, a situated and relational mode of being that precedes the various 'bifurcations' of actuality and nature that characterise much of modern philosophy. Rather than a 'view from nowhere', we are always engaged in a particular embodied situation and related to other things in a structured whole. For Whitehead, we do not simply 'have' a body. Rather, our experience is always *with* our bodies, since 'the *how* of our present experience must conform to the *what* of the past within us' (S 58). In seeing, hearing and touching with our bodies, we are already interconnected with symbols and symbolising relations that are woven into the body-world. Rather than an atomistic collection of bare sensations or pure impressions connected through 'high grade' inferences or mental processes, what we perceive is a perspective on a unified whole.

Whitehead begins his enquiry into the more fundamental symbolism or significance of experience much like Merleau-Ponty, with the perceptual phenomena of the lived body, a 'primacy of perception' that functions as a unifying ground for experience. We are surrounded by our bodies, involved in the world, and situated in the here and now. High-grade mentality, as Whitehead puts it, is not required for our lived experience of the world, but is – as we will see – dependent upon it. The primary type of symbolism moves from sense-perception to bodies and, by beginning with the primacy of perception and the body, Whitehead immediately indicates his intention to problematise those bifurcated models of experience that posit a self-contained subject of experience locked up in the private 'inner' space of the mind, a subject that tries to connect its experience to an external world. Various forms of empiricism, rationalism and idealism have tended to view the problem of perception as a set of ideas in the mind, and the question is how we can know that they accurately 'mirror' nature or reflect the world. For Whitehead the most fundamental kind of symbolism precedes these bifurcations, opening onto an 'undiscriminated background' (S 43) or field that we 'conform' to and create with, allowing us to make sense of things. In seeing, hearing or touching, I make sense of things against this background without constitution or mediation by the mind. As Whitehead famously puts it, we do not dance with sensations and then infer a dance partner. In the symbolising relation, perspectives merge into a unity and acquire their meaning. Whitehead would agree with Heidegger, and contra Kant, that the real scandal is not the lack of proofs for the external world, but that they are attempted over and over again and still expected.[3] The coloured shape is not a 'projection' on the interior surface of the mind which then needs to be reconnected with a world outside, but functions more as an abstraction, effect or objectification of a bodily intentional correlate that conveys or transfers the world into my experiences. As Whitehead puts it, 'my process of "being myself" is my origination from my possession of the world' (PR 81). For the transfer or reference to take place, there must be some common element between the symbol and its meaning, but as Whitehead stresses, there is nothing automatic or anthropogenic about this symbolic movement. Indeed, men and puppy dogs can disregard chairs, and 'a tulip which turns to the light has probably the very minimum of sense-perception' (S 5). In addition, Whitehead says that this symbolic directedness can be mistaken, misleading, or otherwise break down – as Heidegger would say, the hammer or the equipment breaks down – and we avoid the deception only with effort. The error turns us reflectively towards our interpretive and conceptual schemes in order to revise them.

Symbolic Reference and Originary Symbolism

In the basic description given above of the withness of the body, Whitehead introduces and develops a distinction between 'sense-perception' – what he will call 'perception in the mode of presentational immediacy' – and 'causal efficacy'. It is the fusion of these two pure types of perception, the transfer or directedness between them, that Whitehead calls 'symbolic reference'. If causal efficacy gives us 'direct experience' or immediate acquaintance with fact, then symbolic reference gives us a derivative experience that can be trusted only if it satisfies certain criteria exemplified in causal efficacy. These distinctions form the basis for Whitehead's critical inversion of Hume and Kant.

The distinction between presentational immediacy and causal efficacy is discussed throughout this volume, and so I will not elaborate in detail on it here. However, I do want to emphasise the differing modalities of space-time that underlie the distinction. These differing modes of perception point to an 'originary' symbolism, a structure of space-time that conditions all experience and life for Whitehead. Presentational immediacy is bound to the modality of the present, or what Whitehead calls 'the doctrine of simple occurrence' (S 38), whereas causal efficacy rests upon a certain conception of spatio-temporal extension. Simple occurrence is constituted by atomic units of space and time that are fully present to themselves in a moment of non-extended absolute indivisibility. The central problem here is how the units succeed each other in a synthetic relation, since, as non-extended atomic units, only the private immediate attributes are disclosed and, as a consequence, the experience of space, time, memory and identity cannot be grounded in a real world: 'there remains only what Santayana calls "Solipsism of the Present Moment"' (S 33). Mere sense-perception in presentational immediacy does not account for the whole of our experience because it 'gives no information as to the past and the future' (PR 168).[4]

In contrast, perception in the mode of causal efficacy is particularly characteristic of temporal extension. Beyond the simple present, our experience includes temporal and spatial extension expressing the mutual immanence of all actualities. As Whitehead puts it: 'causal efficacy is the hand of the settled past in the formation of the present' (S 50). The present 'conforms' to the past, or later events confirm the presence in them of earlier events, and this conformation of the present to the immediate past is experienced reality. In addition, the present occasion not only shares in the nature of the immediate past, but modifies it and adjusts it in anticipation of the future. As Whitehead puts it, 'the present moment is constituted by the influx of *the other*' (AI 181). Although Kant accepts that causal efficacy is a fact of the phenomenal world, it is not presupposed in the data of perception. Rather, 'it belongs to our ways of thought about the data' (S 37). For Whitehead, Kant's claims here presuppose Hume's 'simple occurrences' or momentary events, which are based on 'the extraordinarily naïve assumption of time as pure succession' (S 34). Pure succession is the notion of time as indivisible moments or units that succeed each other, but do not have any relation to each other. Whitehead compares pure succession to colour, in that there is no 'mere colour' as such, but always some particular colour. Similarly, there is no pure succession, but always some 'particular relational ground' (S 35) in terms of which succession proceeds. Pure succession, for Whitehead, is an abstraction from the fundamental reality of conformation (S 38). Conformation divides the present, opens it to the influx of the other and is at work in every moment from the beginning through causality. Causality is perceivable in the relation of conformation between present and past occasions, and is perhaps most prominent when the organism is 'low grade'. As Whitehead puts it, 'time in the concrete is the conformation of state to state, the later to the earlier; and the pure succession is an abstraction from the irreversible relationship of settled past to derivative present' (S 35). Underneath the adventitious show of presentational immediacy lies the more primitive causal efficacy of our bodily organs and the vague world beyond them. In causal efficacy all organisms are conditioned by the environment, whereas sense-perception is enjoyed mainly by 'advanced organisms'.

Presentational immediacy and causal efficacy are 'pure' modes of perception, but experience synthesises and combines them in the complex mode of symbolic reference. They both, in their differing modes, 'objectify' things in the environment directly, but their complete purity is unobtainable. Independently they are abstractions. As Whitehead says, 'perception in the mode of causal efficacy discloses that the data in the mode of sense-perception are provided by it' (S 53). Thus, in practice, the two modes of perception are fused and interrelated through what Whitehead calls a 'common ground' that enables symbolic reference to take place. For Whitehead, the common ground that enables the 'intersection' between the two modes of perception is 'sense-data' and 'locality' (S 49). Sense-data play a double role: on the one hand they exhibit spatial relations in the present, and on the other they show the immediate past of our bodily organs pressing in on the experience. If we see red, we see with our eyes, and the experience also refers to, or is localised in, an 'external space', say the red of a traffic light. The localisation is clearly demarcated when reference from presentational immediacy is involved, but beyond the bodily organs, reference to causal efficacy shades off into the vague and indefinite. Thus, the common ground of the two modes of perception is a spatio-temporal system in which the past conditions the present and acts as a symbol for the near future.

But this common ground does not necessitate that reference will take place, nor does it determine which element shall function as the symbol and which the meaning. 'There are no components of experience which are only symbols or only meanings' (S 10). Stripped back to its essentials, symbolism is a mode of experience that functions by being directed towards or related to another component or element in the experience. Whitehead sometimes talks about this as the 'vector character' involved in experience. In other words, the causal influences within the symbolising process have a direction which is marked or felt more or less directly by the other elements. As Whitehead puts it, this time in relation to 'feelings': 'feelings are "vectors"; for they feel what is *there* and transform it into what is *here*' (PR 87). At the base of experience we find a symbolising process which requires that a component refer to, or be directed toward, or feel another component for it to be itself, or for it to convey 'meaning'. As Whitehead puts it, 'considered by themselves the symbol and its meaning do not require *either* that there shall be a symbolic reference between the two, *or* that the symbolic reference between the members of the couple should be one way on rather than that other way on' (S 9–10). In classical phenomenology, concepts of intentionality function in terms of a directionality that is one way and unidirectional, a directionality without turns or reciprocity that runs between the knower and the known. For Whitehead the symbolising process is a relation where either term in the relation can play one of the roles, but the directionality remains. In relation to language, Whitehead gives the example of the word 'tree' and the tree itself, and asks why the word 'tree' is a symbol for trees. The tree itself could just as well function as a symbol for the word. Abstracting from human experience, symbolism does not require consciousness, subjectivity or agency as such, only that one entity invoke, respond to, or be 'present in' another, although the relation and the components will vary greatly. Each component enters into the experience as 'equals', with no one component taking precedence over the others. Nothing is simply present or absent, since each element presupposes syntheses or referrals which prevent any one element from simply referring to itself.

This notion that an individual element is present in all of the others for it to 'symbolise' is, I would suggest, the originary symbolism at the heart of Whitehead's metaphysics. No symbol is ever fully present to itself, but refers to the spacing by means of which the symbolising elements are related to each other. No symbol can be meaningful in and of itself, but must be inscribed within a chain of referrals that bears the marks of all these others within itself. All that originary symbolism in Whitehead's sense requires is that there is something to be received – a transmission, passage or transference of the recepta – and an act of reception or inheritance. This passage or transference is the coming of time that divides every moment in advance by relating each instant to something other than itself. However, although the radical implications of this chain of references and referrals provide the rationale for the 'epochal' structure of temporal occasions that comes to the fore in *Process and Reality*, those implications will also, I suggest, provide a challenge to that rationale when viewed through a Deleuzean lens.

The structure of originary symbolism can be traced to the condition of time, since the general function of symbolism is to mediate between past and future. For one element to symbolise another, it must conform to the immediate past and anticipate the immediate future. Symbolising necessarily occupies a duration in which the present is immediately divided by conforming with a past that is preserved in the present and a future that is anticipated, invoked or elicited. Whitehead is very close to William James's famous descriptions of a 'specious present', albeit generalised beyond the stream of consciousness, to indicate that experience never captures the individual present moments of a 'now', but only a present that stretches back into the past and forwards into the future.[5] The present is 'specious' in that it is never immediately available in an instantaneous now-moment, 'knife-edge', or atomic sensation as such, but only in a block or epoch that stretches through a continuity of immediate past and future moments. However, like James, the 'block' or durational act itself for Whitehead is not a continuity; only the moments in the duration are felt continuously. Whitehead not only adopts the phrase 'specious present' and the idea that individual units of experience come in epochs, but also accepts James's view that although the percipient event is temporally extended, the act of perceiving is itself a unity that is unextended and indivisible. In other words, the 'content' of the units of experience or objects of symbolic reference undergoes temporal extension, but the 'form' remains unextended.

As we have seen, for Whitehead symbolic reference combines a spatialising moment which retains the immediate past and anticipates the immediate future, but as a formal whole the experience is given as a unifying epoch or indivisible 'living presence' that does not have temporal extension. This is Whitehead's (and James's) response to Zeno. As Whitehead puts it, 'If we admit that "something becomes", it is easy, by employing Zeno's method, to prove that there can be no continuity of becoming. There is becoming of continuity but no continuity of becoming' (PR 35). Symbolising units or actual occasions become, and they constitute together an extensive world in which only extensiveness becomes, 'but "becoming" is not extensive' (PR 35). Becoming occurs within the symbolising process, but the act of symbolism occurs all at once, so that reality grows for Whitehead, just as it does for James, by 'buds or drops of perception' (PR 68).[6] Upon reflection you can divide the experience analytically, but as it is immediately given it is all or nothing. Thus Whitehead writes:

> the conclusion is that in every act of becoming there is the becoming of something with temporal extension; but that the act itself is not extensive in the sense that it is divisible into earlier and later acts of becoming which correspond to the extensive divisibility of what has become. (PR 69)

Whitehead distinguishes the 'form' of becoming – the structure of the act of experience or symbolising – from the content in which something becomes in order to shore up the infinite regress that Zeno's paradox threatens. The epochal structure of occasions is supposed to put an end to temporal regression by being constitutive of itself and providing a unity and a synthesis to the becoming that mediates reference. The act of becoming, as a non-temporal unity, thereby ensures that the chain of symbolic references does not continue without origin or end.

This structure of time underlying the process of symbolising can be usefully contrasted with Deleuze, since in his accounts – as we will see – the very movement of temporal regress is to be equated with a structure of temporalisation that disrupts seriality, reverses before and after, and overflows the stream of experience. In such conditions we find an originary symbolism expressed as an open structure of differences or a 'becoming unlimited' that precedes and makes possible the phenomena of perception and the symbolising process identified by Whitehead.

Deleuze and Perception

> The task of perception entails pulverizing the world, but also one of spiritualizing its dust. (TF 87)

The accounts of perception scattered throughout Deleuze's works are rich and complex. In several of these texts Deleuze has not hidden his affinity with Whitehead's thought and indeed with Whitehead's understanding of perception, at times explicitly embracing Whitehead's concept of prehension. However, Deleuze's (and Deleuze and Guattari's) own concepts of perception provide an interesting contrast to Whitehead, since they show both striking resemblances and a variation of the process view on originary symbolism. We said that Whitehead's thought is driven by the problem of dualism and the 'bifurcation of nature'. Deleuze is also deeply concerned with this problem. Deleuze's own term for bifurcation is 'transcendence'. Transcendence is the appeal to any idea, being, identity, principle or form that is external, above, beyond or 'other' to the experience it attempts to explain. It is external, beyond or 'other' in the sense that it does not explain the genesis of a phenomenon by its immanent features, but appeals to something external to its genesis, something immutable, ready-made or unchanging to explain a thing's coming to be. For Deleuze the achievement of philosophy would be to break with transcendence to arrive at what he calls 'pure immanence', an immanence of being and thought that cannot be reduced to objective matter or subjective mind. The achievement of immanence here would be to account for the relation between bifurcated terms, not through a transcendent 'third' term or identity, but in terms of a difference internal to and constitutive of the distinction. Pure immanence is immanent only to itself, a difference that differs internally with itself. As soon as immanence is immanent *to* something (other than itself), then difference has become 'externalised', and experience has been bifurcated or become transcendent.

In terms of Deleuzian immanence, to try to understand perception in the context of the traditional epistemological relation between perceiver and perceived is to set out with a badly posed problem, since one starts out from a plane of transcendence where difference is already externalised. Rather, for Deleuze perception is itself a difference, a more fundamental internal difference that is constitutive of any distinction between perceiver and perceived. As we have seen, Whitehead's account of symbolism gets behind the distinction between perceiver and perceived and turns on the importance of the distinction between two pure modes of perception (causal efficacy and presentational immediacy) and their fusion in the mode of perception that Whitehead calls 'symbolic reference' and the role that this plays at the base of all experience. Throughout his work Deleuze also distinguishes (at least) two modes of perception, and emphasises what we can call 'differential efficacy': the efficient power of difference to affect and constitute experience. The Deleuzian differential operates down in the depths of pre-human or non-human experience as a genetic element of perception, a 'molecular', 'microscopic' or unconscious perception. The differential is a 'transcendental' condition of experience, a supra-category of perception itself, that does not subsume objects under a universal generality, but constitutes itself alongside the differences actualised with it. The movement of difference is internal to its effect, a difference from itself that is articulated in two modes. Indeed, in his remarkable book on Leibniz (*The Fold: Leibniz and the Baroque*), Deleuze uses Whitehead's distinction between microscopic process and macroscopic process to understand how differential perceptions are created as folds in the virtual, like so many little creases and pleats requiring a process of differentiation or creation to be unfolded. Microperceptions are little folds that unravel in every direction. They are minute, vague, obscure or confused perceptions, or what Whitehead calls perception in the mode of causal efficacy. They make up our macroperceptions, which are conscious, clear and distinct perceptions, or perceptions in the form of Whiteheadian presentational immediacy. For conscious perception ('high-grade' forms of symbolic reference in Whitehead) to happen, macroperceptions must be continuously destabilised by a series of infinite microperceptions that prepare a new perception. Microperceptions are both constituent elements of perception as well as tiny agents of change and passage that nourish the new perception. The macroscopic distinguishes perceptions and appetitions one from the other so that we arrive at composite folds and forms immediately present. The microscopic, however, issues from a power of causality that can be seen from two perspectives. On the one hand, the microscopic conveys an infinite world that it contains, but, on the other hand, every conscious perception equally implies an infinity of tiny perceptions that prepare, compose or follow from it. The relations between the microscopic and the macroscopic are not to be conceived as part to whole, but as a relation between what is ordinary and what is remarkable. What comes to our attention in perception is made up of elements which are unnoticed. Conscious perception is the product of (at least) two heterogeneous and unconscious elements entering into a differential relation. Out of the obscure and dark background of inconspicuous tiny perceptions – the 'obscure dust of the world' (TF 90), as Deleuze puts it – some perceptions are formed and drawn into clarity.

For example, the colour green: yellow and blue can surely be perceived, but if their perception vanishes by dint of progressive diminution, they enter into differential relation that determines green. And nothing impedes yellow or blue, each on its own account, from being already determined by the differential relation of two colours that we cannot detect, or of two degrees of chiaroscuro (TF 88). Out of the dark and evanescent perceptions (yellow and blue) a clear perception (green) is established. And yellow and blue can also be clear perceptions if they are drawn into clarity by differential relations among other minute and obscure perceptions. Such is the case with Leibniz's other well known examples of hunger, the sound of the sea and the sleeper for whom all of the little creases and folds of the body-world environment enter into relations that produce an attitude and position that induces sleep. If 'good macroscopic form depends upon microscopic processes' (TF 88), this is because the little perceptions that are selected are the ones engaged in differential relations and produce the quality at the given threshold of consciousness. Even the tiniest animals have little glimmers of clarity that filter through the dark expanse of nature that they convey: little glimmers that enable a recognition of food, enemies, perhaps even a partner. Deleuze's favourite example here, taken from the great ethologist Jakob von Uexküll, is the tick. The tick has essentially three clear perceptions: a perception of light, an olfactory perception of its prey and a tactile perception of the best place to burrow. All the rest is a numbness, 'a dust of tiny, dark and scattered perceptions' (TF 92) that make up the vast backdrop of nature.

Thus, as for the Whitehead of *Process and Reality*, macroscopic and microscopic levels of perception together inform the theory of the perceptual Idea or originary symbolism in Deleuze. Symbolism is expressed through an internal method of genesis and a novel theory of relation. Symbolic Ideas inhabit an immanent yet imperceptible realm, and experiments with accessing this realm amount for Deleuze to a 'pedagogy of the senses' that would form a new 'transcendentalism' (DR 237). If the method of genesis owes something to Kantian transcendental conditioning, albeit radically reconfigured, the new concept of relation – 'vice-diction' as Deleuze calls it in *Difference and Repetition* – draws upon a Leibnizian-inspired mathematics and metaphysics of calculus that enables the Idea/symbol to be approached as a *multiplicity* that is always subsuming each thing under its unique 'case'.

In all of Deleuze's discussions of perceptual Ideas or symbols there is a critique of 'natural perception' and its conditions. As with Whitehead, these conditions for Deleuze function as coordinates that serve to anchor a conscious perceiving subject in the world, expressed most famously in the phenomenological 'all consciousness is consciousness of something'. Both Whitehead, as we have seen, and Deleuze are critical of this model of intentionality insofar as it overvalues consciousness at the expense of the body and other symbolising features (backgrounds and depths, schemas and multiplicities, etc.). For Deleuze the reliance upon 'common sense' and the persistence of the model of recognition and representation within natural perception are characteristics of an essential overvaluation of consciousness that distorts the real transcendental conditions of perception. Rather than consciousness with its representational 'states of affairs', a perception in Deleuze, much like in Whitehead, opens onto its own conditions when a body-sensation induces a fundamental encounter with a problematic field:

> What is called 'perception' is no longer a state of affairs but a state of the body as induced by another body and 'affection' is the passage of this state to another state as increase or decrease of potential power through the action of other bodies. Nothing is passive but everything is interaction, even gravity . . . Interaction becomes *communication*.[7]

In *The Logic of Sense*, Deleuze poses the problem of perception and interaction as a problem of knowing how to attain the universal communication of events beyond the contradiction and incompatibility of states of affairs, a problem of how the individual can transcend its form and identity in order to attain this state of communication and be worthy of the power of life that passes through it. We perceive ourselves as the event by grasping the event actualised within us as another individual grafted onto us, 'like a mirror for the condensation of singularities and each world a distance in the mirror'.[8] This is the sense of 'counter-actualising' perception or reversing perspectives, drawing on the potential that cannot be separated from states of affairs and through which they take effect. In the Leibnizian world the potential of this infinite series is subject to rules of exclusion that govern convergences and divergences (one question here will be whether Whitehead's world is also subject to similar rules). Leibniz will not allow all possibles or incompossibles to exist: possible worlds cannot pass into existence if they are incompossible with the world chosen by God. For Leibniz, as for Whitehead and Deleuze, points of view are not points of view on things, but are themselves beings. They are Life or living perspectives. However, for Leibniz only those points of view that converge are open to each other, becoming converging points of view on the same city. For Deleuze a point of view is opened onto a divergence which it affirms, each town or city another point of view, each point of view another town, a point of view on a point of view. Counter-actualisation becomes the affirmation of infinitive distance, and incompossibility becomes a means of communication between diverging points of view.[9] Beneath the form of identity and persons there is an infinite becoming of depth; points of view diverge and are affirmed in their divergence and through their difference. Each series resonates inside the others and returns outside of itself, an exploration of the most distant and the most deep, along lines with multiple branchings. Rather than remaining content with a broadening or generalising of experience in the Whiteheadian sense, Deleuze's 'continuous variation' will go 'beyond the decisive turn'[10] of experience, pushing the concept of perception beyond any 'point of view' or atomic perspective to its virtual limit in what Deleuze and Guattari call *'becoming-imperceptible'*. All becomings (becoming-woman, becoming-animal, etc.) for Deleuze and Guattari are ultimately rushing towards this becoming imperceptible, 'the immanent end of becoming, its cosmic formula' (ATP 279). The 'cosmic formula' of becoming for Deleuze and Guattari would be a process of 'worlding', where nothing is ever 'in itself' since the task is 'becoming everybody/everything' (ATP 280).

The emphasis in the Deleuzian formula of becoming is quite different from Whitehead's, since for Deleuze 'the organism is that which life sets against itself, in order to limit itself' (ATP 503). In other words, for Deleuze the individuality or atomicity of existence is a limitation of life, and in terms of the human it is a limitation and restriction of non-human becomings of the human. Becoming imperceptible is a process of non-human becoming differentially related to the whole field of becomings, expressing itself across the whole continuum of 'non-organic life', at one with the molecular and the cosmic. As such, becoming imperceptible is, to use Deleuze and Guattari's terminology, a 'line of flight' from the actualised and extended individuality of existence, a 'destratification' of its 'molar' form in favour of the relativity of 'molecular perception', a plunge into the pre-individual singularities and non-subjective individuations that traverse the plane of consistency. Becoming imperceptible creates a line that will lead to a 'world without others' beyond any point of view, freeing up singularities, events and intensities, releasing 'percepts' from perceptions and 'affects' from affections.

We can say that, on the one hand, there is 'clear-confused' perception where perceptions or points of view are for Deleuze an actualisation of pure becoming, an arrest of its virtual potential in an organised perceptual form. On the other hand, there is 'distinct-obscure' becoming imperceptible as a task to be assumed, a dissolution or 'overcoming' of point of view in the virtual flow of perceptual becomings. Following Bergson, we might call this latter an 'objective perception', in which the identity of the object as seen by a seeing subject vanishes and is swallowed up in differences that take us beyond the turn of human experience. The subject–object relation begins to break down since 'each point of view must itself be the object, or the object must belong to the point of view' (DR 56), and gives way to a 'perception-image' unanchored from the body, which is without identity, boundaries, limits or human coordinates – a generalised 'view from nowhere', 'any-view-whatever'. However, this last expression risks implying a reductive indiscernibility, an empty identity, a Parmenidean One on an indifferent plane of Being (this has been remarked upon by Badiou and others). On the contrary, the point of view here is a complex of space and time that imposes its own 'scenery', carries its 'phantastical' point of view with it, each time recreated. Rather than this 'view from nowhere', perhaps the best designation for the movement of becoming imperceptible would be Deleuze's own use of '*Erewhon*': 'Following Samuel Butler we discover *Erewhon*, signifying at once the originary "nowhere" and the displaced, disguised, modified and always recreated "here and now"' (DR xxi). Deleuzian erewhon claims to reach the genetic element of all perception, the 'differential of perception itself', the 'virtual point' beyond the 'turn' of experience that is both a differentiation and an integration, a genesis that is also a synthesis. Deleuzian originary symbolism is a becoming imperceptible or differential perception that appears to reach further down than Whiteheadian epochal occasions, tracking back beyond the 'subject' of prehensions with their points of view and following the path to the bend at which '"reason" plunges into the beyond' (DR 282).

Thus the Deleuzian challenge is this: how can we think, feel and live with this plane of pure perception or immanent difference? 'How can we reach this "plane", or rather how can we construct with it, and how can we draw the "line" leading us there' (ATP 503)? This is the task of becoming imperceptible, or what Deleuze and Guattari have alternatively called the problem of 'how to make yourself a body without organs' (ATP 149–66). Thus the problem of becoming imperceptible in Deleuze entails knowing how we move from differential perception to perceptual difference and back again, from anywhere or nowhere to 'here-now' and back again. It is a 'learning' process, as Deleuze says, of 'turn and return', involving movement 'from the cosmological to the microscopic, but also from the microscopic to the macroscopic' (TF 87), or from the 'Event' of perception to the perception of the event ('state of affairs') and back again. For Deleuze, only when perception is drawn into this movement of difference and repetition or 'counter-actualisation' and sufficiently molecularised will it bring forth forces 'attributable only to the Cosmos' (ATP 347).

Between Deleuze and Whitehead: On Becoming and Time

Both Deleuze and Whitehead develop their views of perception in relation to the broader problems that their works seek to engage. For both philosophers, the problem of dualism understood as the bifurcation of nature or transcendence is one of the motivating problems of their work, and both find resources in the empiricist and rationalist traditions for their own ontological accounts of perception. For both Deleuze and Whitehead perception puts us directly in contact with the real; indeed perception *is* the real. Whitehead and Deleuze share a commitment to what I have called an 'originary symbolism', or what we can term a 'primary process' of perception, in which experience directly conditions, prehends or communicates with the thing itself. This more fundamental form of perception is a dynamic and constitutive relation between 'prehensions', a temporal and affective relation of occasions, differences, becomings or 'images' where one flow or series intersects with another, affects it and/or is affected by it. To reach this idea of a primary process or 'pure perception', Whitehead generalises from a human subjective point of view, yet reaches beyond any individual consciousness. As we have seen, however, in Whitehead's metaphysical works the subject–object structure of this process is increasingly atomised, perhaps most especially in *Process and Reality*, at the expense of the 'continuity of becoming'. Deleuze, in contrast, appeals to the continuity of a non-human 'pure perception', an 'originary nowhere' or virtual plenitude, and sees an atomised or individual 'point of view', however extended, as a limitation of the virtual field of differences, a canalisation of the becomings that traverse the non-organic flow of life.

The accounts of symbolism and perception in Whitehead and Deleuze are conditioned by their understanding of time and becoming. Whitehead insists in several texts, especially *Process and Reality*, that time is atomised and epochal. Like James, for Whitehead reality grows in drops and buds, and so time cannot be thought of as a continuity. As Whitehead says, 'temporalization is not another continuous process. It is an atomic succession. Thus time is atomic (i.e. epochal), though what is temporalized is divisible' (SMW 126). Whitehead arrives at this position as a result of an analysis of Zeno. If we analyse the act of becoming with the premises that something becomes and that every act of becoming is divisible into earlier acts of becoming, then we end up in the contradiction of an infinite regress where nothing becomes. To use Whitehead's example, if we take an act of becoming during one second, we can divide that act into two – namely, the act of becoming in the first half of the second, and the act of becoming in the second half of the second. Operating with the above premises, 'that which becomes during the whole second presupposes that which becomes during the first half second. Analogously, that which becomes during the first half second presupposes that which becomes during the first quarter second, and so on indefinitely' (PR 68). If we consider the process of becoming up to the beginning of the second in question and ask what becomes, Whitehead concludes that 'no answer can be given' (PR 68). Infinite regress leads to a contradiction in the notion of becoming because if the act of becoming is itself temporally divisible, it cannot act as a synthetic unity for something to become, but must itself be subject to further acts of becoming. Fundamentally, no symbolising or perceptual process can be self-constituting if it is subject to the temporalisation of pure becoming. Indeed, 'these conclusions are required by the consideration of Zeno's arguments' (PR 68).

For Whitehead the claim that the regress of becoming would make it impossible to think the concept of originary symbolism appears as a presupposition of his mature thought. What is distinctive of Deleuze's process philosophy is the idea that originary symbolism – or the originary 'logic' or donation of sense – not only does not need a primal ground or non-temporal unity to perform the requisite synthesis, but is constituted by a *pure* or *absolute becoming* that functions as a condition for the movement of temporalisation. Infinite regress or 'becoming unlimited' is the originary movement of time in Deleuze, a movement that ungrounds any origin or end. For Deleuze, out of the paradoxes of pure becoming emerges a conception of time consistent with infinite regress and indefinite divisibility.

For Deleuze we can think of originary symbolism as the synthesis of time ('disjunctive synthesis') without positing a non-temporal act of atomic unity. Indeed, all that is required to exemplify originary or arche-symbolism in Deleuze's sense is the structure of symbolic reference (presentational immediacy – past-future) released from its epochal unity and subject to absolute differentiation or becoming. The now of immediacy is divided between its own becoming past and its pointing towards a future. Immediacy is a primal symbolism or 'reference' that enables repetition across time and perception to take place. Like Whitehead's, Deleuze's originary symbolism is conditioned by time, but it is a time that 'ungrounds' all temporal unities in a generalised 'opening' of experience to difference and becoming. One can see this ungrounding in Deleuze's account of the structures of time in his *Difference and Repetition*. Here Deleuze lays out three syntheses of time and gives an account of how they perform this work. What I suggest here is that the conception of time that underpins Whitehead's notion of originary symbolism is consistent with Deleuze's description of time given in the first synthesis, but it does not include anything like the second and third syntheses, since these syntheses cannot be recognised within the successive 'state to state' movements of Whitehead's epochal structure of time. Deleuze offers an account of time and becoming that is prior to and makes possible the modes of perception and symbolism that Whitehead describes.

Each synthesis in Deleuze's account involves processes that function as a perspective from which to view the operations of the other syntheses. Although the syntheses each correspond to the modalities of time (past, present, future), no temporal process is independent of the other. There is no present that does not gather an element of the past and the future, no past in general that does not allow the present to pass, and no future that does not open the present and the past to new events. Each synthesis is determined by a primary modality that directs the processes involved, but each modality operates 'intratemporally', as Deleuze says, such that the syntheses work through each other and upon each other in a moving complex structure of interactions.

In the first synthesis – what Deleuze calls the 'living present' – experience is inscribed through a process of 'contraction'. This is close in some respects to phenomenological accounts of the 'living present' but, more importantly for us, it is close to Whitehead's epochal becoming and James's specious present. Contraction is a 'passive' synthesis whereby the instants in succession are drawn into each other. In contraction, instants or particular elements are pulled into one unified experience. The succession of instants does not in itself constitute time, but indicates 'only its constantly aborted moment of birth' (DR 70). This notion is close to Whitehead's notion of time as 'conformation', where 'pure succession' is an abstraction from a relational ground 'in respect to which the terms succeed each other' (S 35). There are two key points of comparison here that bring Deleuze's account of the first synthesis of time close to Whitehead. Firstly, contraction is passive. In other words, this operation takes place beneath the 'active' functions of the intellect and points to temporal levels and layers of bodily experience that are constitutive. As Deleuze says, it is not a synthesis carried out *by* the mind, but *in* the mind (DR 71). Secondly, the instants in succession are never self-identically present and indivisible among themselves. Rather, each living present is already divided into a past and a future so that the 'now', as a 'knife-edge' moment (to use William James's phrase), is aborted into the extended contraction of the living present. The instants that are contracted can have a greater or smaller duration depending on the individual, the organism and the species.

Again, we could talk about this point in relation to Whitehead or James. In relation to the latter, the point would be that the 'fringe' or 'halo' of the experiences is more or less extended even across the experiences of one individual, and this extension is determined by: the intensity of the contraction; the number, connection and resonance of the instants contracted; and 'the natural contractile range of the contemplative souls involved' (DR 76). The contraction of the living present includes the past by retaining preceding instants and anticipates the future through an expectation of instants that are to come. Deleuze generalises this structure to all living things as a condition for any living organism to have experience (at points Deleuze appears to go beyond what we usually consider 'living' and asks whether it is 'irony' to say that 'everything is contemplation'). Within the living present Deleuze stresses that the 'moments' of the past and the future are not separate and distinct, but are part of the contraction of the living present. Past and future are 'dimensions', as Deleuze says, of the living present. As Deleuze points out, the passive synthesis does not need to go 'outside itself in order to pass from past to future'

(DR 71). The synthesis does not need to go 'outside itself' to pass because it constitutes the dimensions of the past and the future in the contraction. In other words, the living present is already 'outside itself'. The first synthesis is a primary or 'originary' dimension of experience for Deleuze, and it is originary because the subject of the synthesis affects itself and relates to itself as an other by incorporating the dimensions of past and future. In the first synthesis time is 'asymmetrical', meaning that the spatial arrangement and placing of the dimensions (of past and future) are strictly directional, moving from past to future. This is the famous 'arrow of time', vector or succession of phases from earlier to later that, as we have seen, Whitehead insists on in the movement of perception and in his epochs of becoming.

This whole structure of the living present constitutes an originary time without itself being coextensive with the whole of time. The living present is 'intratemporal', and therefore undergoes passage. This constitutes the paradox of the present: the present constitutes time even as it passes within the time constituted. What is it that enables the living present to pass? Here Deleuze moves to a second passive synthesis, deeper than the first. Habit, or the first synthesis, is grounded in memory – not the memory of retention, an active memory that is dependent upon habit, but a deeper memory that constitutes the being of the past. This is not a past that has been present. Indeed, Deleuze suggests that it is 'futile' to try to reconstitute the past from the present. The past is trapped between two presents: the present that it once was and the present that it is now the past of. In other words, Deleuze says we cannot access or get to the past by thinking about it in relation to any kind of present. Deleuze says he is 'unable to believe that the past is constituted after it has been present' (DR 81). This is a key point in the undoing of the 'ordinary' conception of the present, or what Whitehead calls 'pure succession'. The moment is not first present and then passes. Rather, the passage of time is underway from the 'beginning' so that the present is already in a state of passage, having never been present in itself.

So, the present passes and its passage is conditional upon a past that enables the present to become past. However, the past cannot be a past formed from the present. Nor can a past be formed when a new present appears. If a new present were required for the past to be formed, then the former present would never pass. There is, Deleuze argues, a second time in which the first synthesis can occur and receive a ground. The living present requires a second paradoxical dimension for it to pass, but it is a past that has never been present. Memory synthesises a 'pure past' that is contemporaneous with itself. Such a past or time as pure memory is not conceivable in Whitehead's metaphysics of time, since becoming is governed by rules of seriality and succession where the present conforms to a past that was the living present. In addition, Whitehead tends to treat the immediate past as a 'degree' of the present and not really a past at all (and much the same can be said of the future as a degree of the present). The relationship of settled past to derivative present is, Whitehead says, 'irreversible' (S 35). For Whitehead the living or extended present perishes perpetually, and the process by which it becomes, passes and is taken up into new occasions is coextensive with time. For Deleuze the living present is not coextensive with time, but presupposes a pure past, an *a priori* past in general for it to pass. But the pure past does not itself pass.

The coexistence of the living present of habit and pure past of memory are complemented by the third synthesis, which marks a radical departure from what we might call Whitehead's 'rationalisation' of becoming. The third synthesis or 'empty form of time' functions as the structure of the continuity of becoming, or the differentiator for the expression of the other syntheses (present and past). The empty form of time is the 'originary symbolism' for the 'caesura' that divides every moment even as it subsumes the syntheses of present and past. As Deleuze puts it:

> it must be called a symbol by virtue of the unequal parts which it subsumes and draws together, but draws together as unequal parts . . . This symbolic image constitutes the totality of time to the extent that it draws together the caesura and the before and the after. (DR 89)

In other words, we can think of the third synthesis as integrating the past into the present in order to relate to the future, but this way of explicating the empty form of time as an image concerning the totality of time is still only 'introductory' (DR 90). The third synthesis properly concerns the '*eternal return*', which is the key concept that organises the entire structure of Deleuze's *Difference and Repetition* and functions as the thought of the future which subordinates the other stages and leaves them behind. Rather than merely 'drawing off' a difference, as in the mode of habit in the living present, or making difference a variant of a memorial pure past, the eternal return, as repetition of the future, is the 'production of the absolutely different, making it so that repetition is, for itself, difference in itself' (DR 94). Deleuze gives us an image of time without origin or end in which difference in itself repeats and differentiates itself: 'the eternal return has no other sense but this: the absence of any assignable origin, in other words the assignation of difference as the origin which then relates different to different' (DR 125). The only 'in-itself' is difference or pure becoming that functions to unground the other modalities of time summarised as succession and coexistence. The living present and memory of the pure past are conditioned by a deeper order which is not extensive but 'intensive, purely intensive. In other words it is said of difference' (DR 243).

Conclusion: Whitehead, Deleuze and the Process Tradition

For Whitehead a symbolising relation or actual occasion gives perceptual content unity and shows how a stretch of time holds that content together in experience. In other words, the specious present is made possible by a non-extended or momentary unit of becoming that explains an act of experience. Only if there is an actual occasion that holds it together can there be a sensed or perceptual content in time. As we have seen, if the act of becoming itself, not just the content, were temporally extended, then nothing is held together or becomes, and a 'rational' explanation of experience becomes impossible. For Whitehead symbolic reference is a solution to the problem of perception, and the actual occasion is a speculative solution to the problem of temporal experience. Together they show how Whitehead approaches the solution to the problem of bifurcation.

The challenge of Deleuze's process thinking emerges out of a problematisation of this solution. For Deleuze the living present (or the specious present) points to a problem that needs to be explained, a problem that in one way or another informs his work from beginning to end: how is 'sense' made in experience? How is symbolisation possible? If Whitehead's explanation 'rationalises' the problem by positing a non-temporal act (a form of transcendence for Deleuze), Deleuze wants to think about the problem in terms of immanence, and that means the thought of time as absolute becoming. In *Difference and Repetition*, Deleuze pursues an answer in terms of syntheses of time, syntheses that are 'intratemporal' and require explanation through each other. For example, the syntheses of the living present themselves need to be explained by and grounded in a synthesis of memory, a past that has never been present, which in turn is dependent upon a future that will never be present, the repetition of the empty form of time as difference. Deleuze embraces these structures of time because they are required by the thought of pure becoming or difference in itself, where every moment is already divided by a 'caesura' or a 'splitting' such that each moment can be no longer or not yet.

In their philosophical systems both Whitehead and Deleuze are responding to the Heraclitean 'all things flow', that 'ultimate generalization around which we must weave our philosophical system' (PR 208). But each gives a different interpretation of *panta rhei*. For Whitehead the flux is temporally extensive, but – adhering to the logic of some of Zeno's paradoxes – it cannot be absolute because nothing would become. The form or *logos* of the flux corresponds to a non-temporal act such that 'the creature is extensive, but . . . its act of becoming is not extensive' (PR 69). In contrast, Deleuze's interpretation distinguishes him within the modern process tradition as the most recent proponent of flux understood as the affirmation of absolute becoming.[11] As Deleuze remarks, 'we have to reflect for a long time to understand what it means to make an affirmation of becoming' (NP 23). Deleuze says that Heraclitus affirms becoming in two senses. Firstly, Heraclitus affirms that there is no being, only becoming. This is what Deleuze calls Heraclitus' 'working thought' (NP 23). This working thought is consistent with the more conventional understanding of Heraclitean becoming as the idea that both spatio-temporal location and qualities/predicates are subject to change and are perhaps always changing. The working thought is consistent with the idea that 'things' are flows or processes, but flows or processes that are identifiable flows or processes. Secondly, Deleuze claims that Heraclitus affirms the idea that there is a being of becoming, that being is the being of becoming. In what he calls Heraclitus's 'contemplative thought' (NP 23), Deleuze argues that Heraclitus affirms that the being of becoming is return: 'return is the being of that which becomes' (NP 24). This 'contemplative' or speculative thought is Deleuze's notion of absolute becoming, in which – as we saw earlier, with the third synthesis of time – what returns is precisely the difference that does not allow a becoming to be identified. This is the only 'being' that absolute becoming can have. As we have seen, in several of his books Deleuze gives the notion of absolute becoming the Nietzschean name 'eternal return':

> the eternal return is predicated only of becoming and the multiple. It is the law of a world without being, without unity, without identity. Far from *presupposing* the One and the Same, the eternal return

constitutes the only unity of the multiple as such, the only identity of what differs: coming back is the only 'being' of becoming. Consequently, the function of the eternal return as Being is never to identify but to authenticate.[12]

What is authenticated is the 'superior' ever-changing form of what 'is', the transformation of recognised and identifiable values by the originary symbolism of absolute becoming.[13]

Notes

1. The concept of the 'originary' is borrowed from Jacques Derrida to indicate, as he says, an 'irreducibly nonsimple' synthesis of elements that cannot be traced to the traditional concept of origin or absolute beginning (see Derrida, *Margins of Philosophy*, 13–14). In what follows I argue that both Whitehead and Deleuze employ versions of an originary logic in their accounts of perception and symbolism.
2. The notion of power here is an inversion of the famous platonic notion of being as power. The power of the originary emerges from a temporality of becoming which is 'always in a process of becoming and perishing and never really is' (Plato, *Timaeus*, 28a). On this view, nothing has independent existence or exists in itself, or nothing has the power to affect without being affected, because everything is conditioned by the originary alterity of temporal finitude.
3. Heidegger, *Being and Time*, 249.
4. To what extent Whitehead's account of perception is close to phenomenological descriptions of a 'protentive-retentive' structure of internal time consciousness, again generalised, is an interesting question and worth further study. That study would have to begin, of course, with Husserl's *The Phenomenology of Internal Time Consciousness*.
5. See James, *Principles of Psychology*, vol. 1, esp. ch. 9, 'The stream of thought'.
6. James, *Some Problems of Philosophy*, 155
7. Deleuze and Guattari, *What Is Philosophy*, 154.
8. Deleuze, *Logic of Sense*, 178.
9. Deleuze, *Logic of Sense*, 174.
10. Deleuze, *Bergsonism*, 28–9.
11. In his *After Finitude*, Quentin Meillassoux has argued that the principle of non-contradiction must be an 'absolute ontological truth' (71) for any philosophy of becoming, because no entity or process could become other than itself if it already contained its own alterity, as a contradictory entity would. Thus, in becoming 'things must be this, *then* other than this' (70). I think Whitehead would reply that this is less an 'absolute' and more a relative ontological truth for his own philosophy of becoming.

It is relatively true at the level of temporalisation – whereby successive epochs are realised – but in terms of what is temporalised, this 'substantialises' becoming, and depends upon a notion of time that he calls 'pure succession'. Pure succession requires separate independent units or instances that then 'become' something other than what they are. But for Whitehead, as we have seen, this is an abstraction that leaves out the 'particular relational ground' (S 35) wherein the present conforms to the past. Although succession or becoming as a formal process is atomic in Whitehead's sense, what becomes must always be a derivation of state to state, where a present is always divided from itself or is 'perpetually perishing' into a past and a future from the beginning. Whitehead places constraints on his own notion of originary symbolism, on its 'originariness', because in its 'pure', absolute (or Deleuzean) form it violates Zeno's version, derived from Parmenides, of the principle of non-contradiction. On this view, temporalisation would reduce to an infinite regress that converges to nothing, and so 'time would be an irrational notion' (SMW 127). Although I do not have space to deal with this in more detail here, perhaps one of the more important differences between Whitehead and Deleuze rests on the role that the principle of non-contradiction plays in their respective philosophies of becoming. In contrast to Whitehead, who finds a place for the principle in his system, Deleuze presides over a rigorous destruction of the principle as a step towards a rebirth of Heracliteanism.
12 Deleuze, *Desert Islands*, 124.
13 The idea of the eternal return plays a pivotal role in the conceptual economy of Deleuze's *Difference and Repetition*, providing the basis for difference to be given its own concept. Although it requires a detailed separate study in itself, eternal return has at least two key functions worth mentioning here. Firstly, it has a function of selection: the eternal return eliminates all of the negative, the identical, the same. All these forms prevent the thought of difference in itself. Secondly, eternal return performs the function of 'ungrounding', that is, the elimination of the distinction between a ground and the grounded, a foundation and what it founds. Ungrounding leaves one in the world of simulacra, caves within caves. Interestingly, it is modern works of art that suggest this latter idea to Deleuze.

Bibliography

Deleuze, Gilles, *Bergsonism* (New York: Zone Books, 1988).
Deleuze, Gilles, *Cinema 2: The Time Image* (London: Athlone Books, 1989).
Deleuze, Gilles, *Desert Islands and Other Texts 1953–1974* (New York: Semiotext(e), 2004).
Deleuze, Gilles, *Difference and Repetition* (London: Athlone Press, 1994).

Deleuze, Gilles, *The Fold: Leibniz and the Baroque* (London: Athlone Press, 1993).
Deleuze, Gilles, *Foucault* (London: Athlone Press, 1988).
Deleuze, Gilles, *Logic of Sense* (New York: Columbia University Press, 1990).
Deleuze, Gilles, *Nietzsche and Philosophy* (New York: Columbia University Press, 1983).
Deleuze, Gilles, 'A philosophical concept . . .', in Eduardo Cadava, Peter Connor and Jean-Luc Nancy (eds), *Who Comes After the Subject?* (New York: Routledge, 1991), pp. 94–5.
Deleuze, Gilles, and Felix Guattari, *Anti-Oedipus* (London: Athlone Press, 1984).
Deleuze, Gilles, and Felix Guattari, *A Thousand Plateaus* (London: Athlone Press, 1988).
Deleuze, Gilles, and Felix Guattari, *What Is Philosophy?* (London: Verso, 1994).
Derrida, Jacques, *Margins of Philosophy*, trans. Alan Bass (Chicago: University of Chicago Press, 1982).
Heidegger, Martin, *Being and Time* (Oxford: Basil Blackwell, 1962).
Husserl, Edmund, *The Phenomenology of Internal Time Consciousness*, trans. James Churchill (The Hague: Martinus Nijhoff, 1964).
James, William, *Principles of Psychology* (New York: Dover Publications, [1890] 1950).
James, William, *Some Problems of Philosophy* (Omaha: University of Nebraska Press, 1996).
Meillassoux, Quentin, *After Finitude: An Essay on the Necessity of Contingency*, trans. Ray Brassier (New York: Continuum, 2008).
Whitehead, Alfred North, *Adventures of Ideas* (New York: The Free Press, [1933] 1961).
Whitehead, Alfred North, *An Enquiry Concerning the Principles of Natural Knowledge* (New York: Dover Publications, [1919] 1982).
Whitehead, Alfred North, *Essays in Science and Philosophy* (New York: Philosophical Library, 1947).
Whitehead, Alfred North, *Modes of Thought* (New York: Free Press, [1938] 1966).
Whitehead, Alfred North, *Process and Reality: An Essay in Cosmology*, corrected edition, ed. David Ray Griffin and Donald W. Sherburne (New York: The Free Press, [1929] 1978).
Whitehead, Alfred North, *Science and the Modern World* (New York: The Free Press, [1925] 1967).
Whitehead, Alfred North, *Symbolism: Its Meaning and Effect* (New York: Capricorn Books, [1927] 1955).
Whitehead, Alfred North, *The Concept of Nature* (Cambridge: Cambridge University Press, 1920).

3

Uniting Earth to the Blue of Heaven Above: Strange Attractors in Whitehead's *Symbolism*

ROLAND FABER

Strange Attractions

Symbolism is maybe one of the most obscure books of Whitehead's oeuvre: in between grand projects, small in appearance, seemingly integrated with other works, lesser known, and, to a certain extent, considered superfluous. Yet, on second thought, it may be the case that in its fringe existence *Symbolism* holds some gems to be rediscovered and cherished.

Strangely, this is the only book in which Whitehead directly addresses political philosophy; what is more, he develops it from his theory of perception, of all things. It is also the only book foregoing any reference to religion or God, at least explicitly. And it is a book in which all of the elements of thought developed relate directly to thinking and 'articulating' the body physiologically, socially and ecologically, but all fractured through the highly creative concept of a threefold modality of perception, or, in other words, the symbolisation of existence in the evolutionary development of organisms. We are left with warnings of survival, but paradoxically – mediated through our ability to symbolise the world and the future – by developing cultures of ecology.

In elaborating these connections, *Symbolism* can be read in different ways: one way to look at it is to integrate it back into *Process and Reality*, which will become its context; another way would be to view it as expressing the pivotal point of Whitehead's metaphysical work being a metaphysics of experience; yet another way would be to consider it as harbouring a series of strange attractors, which, while not absent from other works, might be found to be more densely interwoven here than elsewhere. This is the approach I am choosing here (not to the exclusion of the other ways that I have explored in other contexts, as have many others).[1]

A 'strange attractor' is not strange because that which it reveals in a series of circumambulations is foreign to, or outside of, any expectation, but because it is somehow surprising in its connectivity, a novelty without apparent system of integration.[2] I look at the text in concert with the symbolism of Deleuze's conceptual movement of infinite speed.[3] Enfolded in the aura of Whitehead's symbolisations, we might escape simplifications of looking at Whitehead's later work as simply prolonging the metaphysical age beyond its death.[4] Instead, I intend to recognise more closely some of the decentring 'attractive' qualities of *Symbolism* that might also shed light on his work as a whole in a slightly *off* manner.

The title I chose is a case in point. It is taken from the *dedication* of the book, where Whitehead relates his thought, once more, to romanticism (not long after *Science and the Modern World* explored the poetic nature of events) and with the sentiment of a connectivity of the impossible as Whitehead meditates with 'unbroken happiness' and 'suffering' upon the world in *this* symbolism: 'a silver thread uniting earth to the blue of heaven above' (S v).

Isn't it 'strange' that this symbolism tucks together an organism (earth) with an eternal object (blue) of a poetic sky (heaven), which not all would accept as a mode of actuality, and that it invokes the 'memorial to Washington' in the poetic nimbus of a 'silver thread'? What is that: a phallocentric lapse, some might say; colour symbolism, others may fear (silver, brown, blue as, e.g., the state flag of Iowa); a Maypole dance, folk researchers may associate? But think of it: a thread unites earth and the colour blue? A strange attractor, indeed! Let us now follow some other strange attractors, sixteen quotes stitched together into three folds: on categories, on perception and on organisation.

On Categories

1. The word 'Caesar'

The word 'Caesar' may mean 'Caesar in some one occasion of his existence': this is the most concrete of all the meanings. The word 'Caesar' may mean 'the historic route of Caesar's life from his Caesarian birth to his Caesarian assassination'. The word 'Caesar' may mean 'the common form, or pattern, repeated in each occasion of Caesar's life'. You may legitimately choose any one of these meanings; but when you have made your choice, you must in that context stick to it. (S 27–8)

The 'doctrine' this quote exposes might be immediately understandable to all who are knowledgeable about Whitehead's thought patterns.[5] But why Caesar? Maybe because Whitehead was reading Edward Gibbon's *The Decline and Fall of the Roman Empire* (1776), which entertains the thesis that it was the adoption of Christianity in the fourth century that contributed to the fall of the Roman Empire – as Whitehead seems to imply when he writes about the 'disruptive tendency due to novelties' that is 'illustrated by the effect of Christianity on the stability of the Roman Empire' (S 70). I will return to this later. Or is it a faint precursor of Whitehead's contention in *Process and Reality* that religion, if it is built on imperial rule, is nothing but its very image (PR 342)?

Yet, in its immediate sense, it takes notice of the fragility of the life of this person, whose name is a symbol in itself, used to craft the archetypical concept of an Emperor/Czar/Kaiser. What is more, Whitehead indicates that whatever we might think of the character of Caesar, and even the life history of all events creating the image of Caesar in his life and for posterity, it is, in fact, the singular events of his life, filled with all the intrigue and the plotting that moved worlds, that in their very fragility are the most concrete mode in which we can assume what it means to be Caesar. In a sense, although all three ways look at any concrete existence, especially if it is living, be it important or not, influential or evanescent, all are evanescent in any moment of their existence – only a fragile synthesis of influences and expectations, errors and desires, plans and interrupted plots. This is *us*, and everyone: only occasions of evanescence; or, as Whitehead poetically proposes in *Science and the Modern World*, 'human life is a flash of occasional enjoyments lighting up a mass of pain and misery, a bagatelle of transient experience' (SMW 192). *Ecce homo*! And that 'character', although abstractive, is a product of conduct: action, self-collection, virtue, if you like. Yet it is not simply abstract, maybe indicated by its function of recognising a friend, as in *Religion in the Making* (RM 61). And potentially we will be offered a glimpse not only of the cover of the historical route of occasions, but also of their collective complexity, the hidden massiveness of influences the route is composed of and that, in infinite patience or pain, have been forgotten and abstracted from. Somewhat like Butler's 'opaque subject', this multiplicity can never tell its story with any completeness or only from recollection, and this weakness is itself already a reference to a new perceptivity, interwovenness, and recollection of any subject from a complex stream of occasions as its existential and ethical situation.[6]

However, as Whitehead's emphasis is on the concrete, which – as we know – is always a concrescence, the recognition of influences happens as an *activity of synthetisation* that does not completely disappear under the weight of constellations and accumulations, projections and postulations, self-redefinitions and excuses: it always, in all its hiddenness, contributes *itself* to the abstractions it entertains and hosts, even if the image of the world indicates the reverse.

2. *The wall contributes itself (Butler)*

This so-called 'wall', disclosed in the pure mode of presentational immediacy, contributes itself to our experience only under the guise of spatial extension, combined with spatial perspective, and combined with sense-data which in this example reduce to colour alone.

> I say that the wall contributes itself under this guise, in preference to saying that it contributes these universal characters in combination. For *the characters are combined by their exposition of one thing in a common world including ourselves, that one thing which I call the 'wall'*. Our perception is not confined to universal characters; we do not perceive disembodied colour or disembodied extensiveness: we perceive the wall's colour and extensiveness. The experienced fact is 'colour away on the wall for us'. (S 15; emphasis added)

The experienced is not a 'dis-essentialised' collection of dislodged attributes, but rather a collection of attributes *localised* externally to, and far from, us. Yet the experienced is never reached in itself since it is only *for us*, as we are constituted by its hidden series of events that in their togetherness with that of the wall are the world *out of which this setting apart* happens. Earlier on, in *Science and the Modern World*, Whitehead has called this hide-and-seek game 'the prehensive unifications of modal presences' (SMW 65), countering any simple location exemplifying his famous fallacy of misplaced concreteness.

It is this wider, always partly hidden *nexus*, in which attributes are inhering – not the collection of attributes, but their complex and distributed *site* – inexplicable by attribution, but exuding their togetherness, which Judith Butler in her recent article on Whitehead 'On this occasion . . .', in the book *Butler on Whitehead*, recognises as evoking the causal past *and* the current synthetic activity of the prehensive historical series.[7] And it appears to be the counter-measure to the danger of non-essentialism. While essentialism can find a substrate in which attributes can inhere, non-essentialism is in danger of losing – along with the essence – the activity that 'does'[8] the attributions as contingent associations, ending in free-floating attributions without any activity, concreteness, drive or contingent connectedness. Or as Whitehead states: 'All real togetherness is togetherness in the formal constitution of an actuality' (PR 32).

Conversely, it seems that Richard Rorty, ruminating about the current-ness of Whitehead, is of this opinion: the difference between abstract and concrete, then, becomes not complicated, but superfluous and irrelevant.[9]

Not so Butler: the 'wall' is not a presupposed substance, nor is it an ossified fantasia of a bundle of attributes; the 'wall' is not a field of free-floating projections, but rather 'a relative point of convergence among culturally and historically specific sets of relations'.[10] The historic route of the 'wall' is not a mere cause or effect (as no event, occasion or society in Whitehead is alone), but the nexic oscillation by which 'synthesis and analysis require each other' (S 26). In other words: the activities of construction and deconstruction are not enemies of concreteness, but its very nature, by which any abstraction is contingently situated and effective, as well as the effect of the very act of effecting. With Nietzsche and Foucault, Butler calls the recovery of this Whiteheadian 'historical route' of occasions (remember Caesar!) 'genealogy',[11] derived from a process of the rediscovery of the activities of 'apprehensions'.[12] The 'wall' has become 'inscribed', but it is no mere site of inscription, nor a mere effect of forces of abstraction disguising themselves as its root, but by actively 'adding itself' to this dramatic game, it is its *process* – an 'addition' not totally defined by the discourse on it, somewhat like Deleuze's AND.[13] And being *patient* of this charade of mutual objectifications, limitations, subjections and masking, it is the fate of relationality or *khora*, as Whitehead muses in *Adventures of Ideas* (AI 134) and *The Principle of Relativity* (R, ch. 2).[14]

3. Abstraction expresses nature's mode of interaction

Thus 'objectification' itself is abstraction; since no actual thing is 'objectified' in its 'formal' completeness. Abstraction expresses nature's mode of interaction and is not merely mental. When it abstracts, thought is merely conforming to nature – or rather, it is exhibiting itself as an element in nature. Synthesis and analysis require each other. (S 25–6)

In a conservative reading of Whitehead's scheme, 'objectification' is the loss of subjectivity or creativity of an actual happening as it passes itself on beyond itself to others of its future.[15] But read in the context of the inscription and performative upsetting of its own 'substantive' regulation, as Butler would say,[16] or given the insight of Deleuze that true return is that of novelty,[17] 'abstraction' is not the loss of performativity or novelty, but becomes the means of the *relationality* of creative activity[18] that creates itself as a route or instigates a mutable character of this route by which it remains available *as* a process of novelty. Hence, although some would conclude that abstraction is either the loss of concreteness and originality or, even worse, the enemy of liberation,[19] the very means of suppression and prejudice,[20] here, in Whitehead's text, it is *abstraction* that cuts through the dichotomies,[21] some would say binaries, that are responsible for the deconstructive analysis of various current philosophies.[22]

As a means of relativity, abstraction undermines the binaries of nature and culture, active and passive, substance and site. In fact, as being essentially an ingredient of any actual happening forming the connectivity of a route or its character, but also undermining it at the same time, abstraction as objectification is the *link* by which nature and culture cannot be set apart either by ontology or by a projective law. *Binaries, not abstractions*, are the enemies of synthesis and analysis, as they hold them apart.[23] In fact, as the next quote demonstrates, for Whitehead, even the hard-wired binaries of mind and matter are pure convention.

4. *A matter of pure convention*

It is a matter of pure convention as to which of our experiential activities we term mental and which physical. Personally I prefer to restrict mentality to those experiential activities which include concepts in addition to percepts. But much of our perception is due to the enhanced subtlety arising from a concurrent conceptual analysis. Thus in fact there is no proper line to be drawn between the physical and the mental constitution of experience. (S 20)

And the same is true for matter and form, so prominently held apart by platonic modes of thought; so impressively opposed in gnostic traditions; and so jealously guarded in their separation with religious sentiments of ultimacy. Instead, says Whitehead:

5. Pure potentiality awaiting (Rorty)

> Aristotle conceived 'matter' – [hyle] – as being pure potentiality awaiting the incoming of form in order to become actual . . . The notion of 'pure potentiality' here takes the place of Aristotle's 'matter', and 'natural potentiality' is 'matter' with that given imposition of form from which each actual thing arises. (S 36)

So, even as matter and form are *modes* of potentiality, 'potentiality' already hints towards another binary disappearing before our eyes: that of possibility and actuality.[24] Instead of their binary functioning in a hierarchy of classifications and unilateral images, altogether harbouring the sentiments against the political implications they offer, they now form only a conceptual instrument of objectification that is always concretely together in the process of synthesis, uniquely so, and in their analysis (or deconstruction) meant only to propel the motive forces of actualisation.

At this point, it is interesting to mention that Rorty, in one of his works as Hartshorne's student, claimed that since matter is potentiality, Whitehead's primordial nature – the synthesis of pure potentials, as would also be true for any event – is *material*, not, as the platonist expectation claims, like a reflex, mentality divorced from the *me on* (that which should not be) of the world of matter.[25] Similarly, Deleuze's reading of virtual events does not allow for the binary opposition of form and matter, potential and activity![26]

On Perception

In a similar fashion, seeking its strange attractions, the following quotes will highlight implications of Whitehead's theory of threefold perception, differentiating causal efficacy, presentational immediacy and symbolic reference. In short: Whitehead analyses the philosophic tradition as having accepted only presentational immediacy as the foundational (or sole) mode of perception, while reconstructing causal efficacy from it (if at all), while Whitehead *reverses* the order of appearance and relegates the most important function to the relation between these two modes as the origin of symbolisation. The details can be found elsewhere and are a matter of discussion.[27]

What I am interested in here is how these modes of perception and their coordination, which are the major subject of *Symbolism*, employ the vanishing of binary stratification in favour of 'mutual immanence', as Whitehead will call this procedure in *Adventures of Ideas*, or in *Symbolism* 'mutual requirement' (S 26).[28] While it may be true that in order to upset the order of stratification Whitehead seems to merely reverse it, he does not actually create a reverse binary, since both modes, in their difference, constitute (and as a 'mutual requirement' in a sense are also constituted by) the third mode, the *synthesis of negotiation* between them (S 8). While Whitehead insists on an evolution of the second mode of presentational immediacy (S 25) that tradition has found to be foundational (S 79), he does so in order to demonstrate the *contingency* and (power-inflicted) *arbitrariness* of the seeming foundation upon which the whole philosophic tradition has based its understanding not only of perception, but of nature – the binaries of matter and mind, form and potential, nature and culture, causality and contemporariness, indeed the spatialisation and temporalisation of the world.[29]

As I will not repeat in detail here,[30] Whitehead's reversal, by its integrating into a process of actualisation as such – that is, actualisation as a process of symbolisation, and symbolisation as an activity of negotiation – also does *not* fall into the traps of reversed forms of post-structuralist modifications of this scheme, as in Kristeva and Irigaray, in which the poetic materiality of causal efficacy or its pre-symbolic nature was highlighted, not without problems of a new stratification.[31]

I will rather repeat here, however, that it seems to me that Whitehead has, with his differentiation of modes of perception, captured Derrida's insight of a metaphysics of presence and its nemesis: *différance*.[32] Presentational immediacy, indeed, functions as the a-temporal, non-spatialised, non-differed and non-deferred 'solipsism of the present moment' (S 29) that constitutes metaphysics of presence. This is also the reason why Whitehead's causal efficacy resonates so well with Kristeva's and Irigaray's unpresentable and un-presencable realm of pre-symbolic and poetic materiality that is not yet contaminated with the phallocentric law of patriarchy.[33] But Whitehead also escapes Butler's charge of essentialism one could see to rest in these approaches to causal efficacy[34] by finding the activity of existence always engaged in a negotiation between them.[35]

6. All in all (Derrida I)

> Again a vivid enjoyment of immediate sense-data notoriously inhibits apprehension of the relevance of the future. The present moment is then all in all. In our consciousness it approximates to 'simple occurrence'. (S 42)

Instead of Derrida's alleged impossibility of 'presence' or the 'present' as mere self-contained absolute,[36] Whitehead understands the illusion of unlimited extension that the present moment may present neither as experience of static eternity nor as illusion per se. Instead it is a mode of *synthesis* that excludes temporality, causality and history in favour of mere occurrence in the now, but is, nevertheless, a highly abstract *modification* of synthesising causal influences gathering to a moment of activity or active togetherness. And it is, in this moment, *real as such*, as in it 'the world discloses itself to be a community of actual things, which are actual in the same sense as we are' (S 21). And as 'medium of intercommunication' (AI 134), the All in All here – not unlike Deleuze's Omnitudo, the All-One[37] – shines in its infinite extension.[38]

Presentational immediacy feels like all in all. As this represents the notion of the Eschaton in Pauline theology (1 Cor. 15: 28) and, more generally, a pantheistic outlook,[39] Whitehead's scheme validates these experiences, not relegating them to a bad or evil or vicious obstruction, but only to an overestimation of their validity given the context of negotiation in which they appear. 'Oceanic feelings' are real, but they are *incomplete* elements of a process of negotiation of time and space and routes of occurrences.

What is more: in doing so, Whitehead has found an interesting way to propel his relativisation of binary structures into the discussion of any process of symbolisation, be it language or architecture or dogs.

7. The symbol and its meaning (Derrida II)

> Considered by themselves the symbol and its meaning do not require either that there shall be a symbolic reference between the two, or that the symbolic reference between the members of the couple should be one way on rather than the other way on. The nature of their relationship does not in itself determine which is symbol and which is meaning. There are no components of experience which are only symbols or only meanings. (S 9–10)

In symbolic reference, symbol and signification are arbitrary.[40] What more can be said? And since this is the case, the problem of truth is now a limited problem within the process of symbolisation. As the negotiation may have different outcomes, it will become *interpretation* (again not only as human activity) and lead to creative activity as synthesis indicating novelty by referential 'error'.[41]

8. *Aesop's dog was a poor thinker (truth as synthetic activity)*

> We all know Aesop's fable of the dog who dropped a piece of meat to grasp at its reflection in the water. We must not, however, judge too severely of error. In the initial stages of mental progress, error in symbolic reference is the discipline which promotes imaginative freedom. Aesop's dog lost his meat, but he gained a step on the road towards a free imagination. (S 19)

Indeed, 'error' is not so bad after all. Since the question of truth here occurs only because error is possible, as 'truth and error dwell in the world by reason of synthesis' (S 21) in which negotiations may fail either to correspond to impulses or to cohere to ideals of harmony, novelty is not popping out of nothing, but out of the *incompleteness* of the process of objectification: 'How that past perishes, the future becomes' (AI 238). Since this is not a matter of 'conceptual analysis', but of the *synthesis* of symbolic reference, 'error' becomes the gate 'that promotes imaginative freedom' (S 19).

At the same time, Whitehead insists that no immediate experience in the modes of 'direct recognition' can err (S 7). Yet, since it is the unavoidable (although at times rudimentary) mutual reference of the causal and the a-temporal that creates differences of interpretation, no experience in any full sense can in the end be uninterpreted.[42] Error becomes another word for inexhaustibility of referential modes yet unexplored. But not everyone might agree with this interpretation (or may see it as error).

In any case, insofar as symbol and meaning are not essentialised by pinning them to either mode of perception or activity of synthesis, in the *analysis* of this process of synthesis, which is itself a mode of synthesis, we will also not discover any signification that harbours essential significance of any binary or non-arbitrary structure.[43] And further:

9. *The pathos that haunts the world (emptiness and significance)*

The contrast between the comparative emptiness of Presentational Immediacy and the deep significance disclosed by Causal Efficacy is at the root of the pathos which haunts the world . . . But sometimes [human beings] are overstrained by their undivided attention to the causal elements in the nature of things. Then in some tired moment there comes a sudden relaxation, and the mere presentational side of the world overwhelms with the sense of its emptiness. As William Pitt, the Prime Minister of England through the darkest period of the French Revolutionary wars, lay on his death-bed at England's worst moment in that struggle, he was heard to murmur, 'What shades we are, what shadows we pursue . . .'. His mind had suddenly lost the sense of causal efficacy, and was illuminated by the remembrance of the intensity of emotion, which had enveloped his life, in its comparison with the barren emptiness of the world passing in sense-presentation. (S 47–9)

Two things become obvious in this quote. First, significance is relegated to causal efficacy, the realm of materiality and poetry, of causality and seriality; it is not, as in Derrida, linguistically related to binaries of mind and matter, forms and potentials, divided into a realm of signification and a realm of significances.[44] And the realm of causal efficacy is also not cut off from significance, which is relegated to a symbolic life 'under the symbolic Law' as the poetic rendering of 'semiotic' *khora* in Kristeva suggests.[45] Rather, significance comes with the causal relationality between events of which 'objectification' is already a part; hence, it is not its binary opposite inducing significances.

Second, the breakdown of the symbolic negotiations can lead to modes of consciousness, or altered states of consciousness, in which, in relation to causal efficacy, the all in all of presentational immediacy induces a deep and expansive feeling of *insignificance*, of not-signification. It must be mentioned here that this feeling, or, better, consciousness (as it is a subjective form of feeling), is a high point of Buddhist consciousness training: *vipassana*, the letting go of any signification process as the appearance of wisdom.[46] Non-essentialism pure! And this is also visible in Whitehead's rendering of Hume's non-essentialist understanding of the relation of occurrences; one is tempted to say with Buddhist philosophy *dharmas*.[47]

10. *A notion of substance, which I do not entertain (Hume)*

> 'The idea of substance must, therefore, be derived from an impression of reflection, if it really exist. But the impressions of reflection resolve themselves into our passions and emotions; none of which can possibly represent a substance. We have, therefore, no idea of substance, distinct from that of a collection of particular qualities, nor have we any other meaning when we either talk or reason concerning it'. This passage [of Hume] is concerned with a notion of 'substance', which I do not entertain. Thus it only indirectly controverts my position. (S 34)

As both Whitehead and Hume are non-essentialists, they differ not in that occurrences are just that, but in their estimation of the context in which this occurs. While for Hume the Buddhist *dharma* analysis might fit well, for Whitehead what is problematic here is the potential non-essentialist fallacy of free-floating attributes or occurrences without any connectivity, which is also, to be clear, not the Buddhist wisdom found in *vipassana*, since in it all is disclosed as related in *pratitya-samutpada*.[48] Hence, Whitehead's non-essentialism – of that 'self-adding wall' – is about the *activity of synthesis* that creates, sustains or destroys organisms by processes of symbolic negotiations. This is the issue to which I will turn in the next section, collecting references from *Symbolism*.

On Organisation

11. *Various organs of our bodies*

> Our most immediate environment is constituted by the various organs of our own bodies, our more remote environment is the physical world in the neighbourhood. But the word 'environment' means those other actual things, which are 'objectified' in some important way so as to form component elements in our individual experience. (S 17–18)

Those 'objectifications', those syntheses, could – since we cannot know of any level of existence in which the complexity of the modes of perception has not been present in its nucleus – well be presumed to indicate already-present processes of symbolic negotiation, symbolic reference and symbolic transference.[49] In other words: bodies *as* bodies are complex multiplicities of mutual symbolic syntheses. As abstraction is not absent from constituting any event, be it nature or culture, symbolic negotiation is not absent either.

Sure, one could again refer to the fact that Whitehead was careful to imply an evolution of perceptive modes so that one could claim that presentational immediacy is the outcome of the organisation of higher organisms, as he seems to suggest in certain passages (S 23). But actually, this is not anything he claims to know with any certainty. It is rather an evolution of the *importance* the development of presentational immediacy assumes in higher organisms that he seems to intend (S 63). Yet we do not have to presuppose that the different modes should be thought to be totally reducible to one or the other at any level of organisation. In fact, Whitehead seems to imply this mutual irreducibility with his 'doctrine of the two types of objectification, namely, (a) "causal objectification", and (b) "presentational objectification"' (PR 58) in *Process and Reality*, and with the cryptic remark in *Symbolism* that 'symbolic reference is the active synthetic element contributed by the nature of the percipient' as part of the process of 'the self-production of an occasion of actual existence' (S 8). Most clearly, this mutuality and the irreducibility of the mode of symbolic negotiation is highlighted when Whitehead speaks of 'symbolic reference' as 'one primitive form of synthetic activity whereby what is actual arises from its given phases' (S 21).

In any case, as bodies employ symbolism by which they are constituted *as* bodies, no body can escape the dialectic, or contrast, between the symbolic and the pre-symbolic. In contrast, however, to Derrida, Irigaray, Kristeva and others, for whom the two realms are in opposition to one another – one the realm of signification, the patriarchal law and 'presence'; the other the realm of pre-signified materiality, maternity and *différance* – Whitehead's pre-symbolic reality of bodies is *in itself* the symbolic play of differences *between* materiality and mentality, or time- and space-related complexity. Again: for Whitehead, 'presence' is *pre*-symbolic, and the symbolic is a *relation within* the pre-symbolic.[50]

Finally, if this is all so, we can also say that symbolic negotiations in the processes of mutual synthesis, which constitute the organisation of bodies, play the role of, and manifest, *this* fundamentally inherent difference (at least in this cosmos and in terms of Whitehead's philosophy of organism): that the cellular structure by which we can differentiate between external and internal processes, external environment and internal organs, is based on the very difference of causal-temporal and presentational-spatial objectification; and that symbolic negotiation is, on a very fundamental level, the expression of the organisation of temporal and spatial structures themselves – again carrying the signatures of the interconnection between materiality and mentality, externality and inwardness, organic difference and unity, as well as causality and presence, all as an expression of symbolisation.[51] This is expressed in the next quote.

12. *From without, from within*

> The bonds of causal efficacy arise from without us. They disclose the character of the world from which we issue, an inescapable condition round which we shape ourselves. The bonds of presentational immediacy arise from within us, and are subject to intensifications and inhibitions and diversions according as we accept their challenge or reject it . . . Our experience arises out of the past: it enriches with emotion and purpose its presentation of the contemporary world: and it bequeaths its character to the future, in the guise of an effective element forever adding to, or subtracting from, the richness of the world. (S 58–9)

Hence, the evolution of the world of organisms is a process of 'mutual symbolic reference' (S 49) and, in the best case, of the 'mutually intensifying' (S 85–6) interactions they support.

13. *The world is a community of organisms*

> The world is a community of organisms; these organisms in the mass determine the environmental influence on any one of them; there can only be a persistent community of persistent organisms when the environmental influence in the shape of instinct is favourable to the survival of the individuals. Thus the community as an environment is responsible for the survival of the separate individuals which compose it; and these separate individuals are responsible for their contributions to the environment. Electrons and molecules survive because they satisfy this primary law for a stable order of nature in connection with given societies of organisms. (S 79)

On the level of communities of organisation, the same mutuality of inside and outside, causality and presentational unity, materiality and mentality prevails. Both on the level of perceptual relations as well as that of organismic organisation, the principle of *mutual immanence* is basic. As developed in different contexts by Whitehead, but especially in *Adventures of Ideas* (AI 134, 167–9, 201), and overlapping with his rendering of the platonic *khora*,[52] mutual immanence has always been to me and in my own work on Whitehead over the last fifteen years the real expression of ultimate reality.[53] Not creativity nor God, not world nor possibilities (RM 90), nor any other notion can, in my understanding, live up to 'the principle without principle' in Whitehead, or as he says in *Process and Reality*: that by which the 'ultimate notions of "production of novelty" and of "concrete togetherness" are inexplicable either in terms of higher universals or in terms of the components participating in the concrescence' so that the 'sole appeal is to intuition' (PR 21–2) of this mutuality.

And it is on the level of ecological relatedness that this principle shows one of its most interesting counter-intuitive characteristics: that of mutual *transcendence*![54] Explored in many contexts, it appears to be the basic way in which Whitehead's organic thought addresses and secures a never-failing, stubbornly irreducible energy between internality and externality, temporality and spatiality, solidarity and privacy, individuality and society, organism and environment, as an expression of the dialectic of interdependence and independence. Here Whitehead is not in agreement with holistic concepts of inter-relatedness; on the contrary, such holism is already constituted and upset by its inverse movement of the 'vast causal independence of contemporary occasions' as the 'preservative of the elbow-room within the Universe', which 'provides each actuality with a welcome environment for irresponsibility' (AI 195).

It is this disrupted connectivity, never facilitating a closure of system or a stagnation of processes of becoming into fixed states of being, which in Whitehead's understanding is the harbinger of novelty, creative advance, the unprecedented, the in/finite[55] processuality of becoming.[56] It is, on the other hand, also the harbinger of chaos, destruction, transformation, and doom, whether viewed from the perspective of evolution or that of social revolution. Yet, in any case, *survival*, in this understanding, is a matter of *symbolization*!

14. Miracles of sensitiveness

> Thus mankind by means of its elaborate system of symbolic transference can achieve miracles of sensitiveness to a distant environment, and to a problematic future. But it pays the penalty, by reason of the dangerous fact that each symbolic transference may involve an arbitrary imputation of unsuitable characters. It is not true, that the mere workings of nature in any particular organism are in all respects favorable either to the existence of that organism, or to its happiness, or to the progress of the society in which the organism finds itself.

> The melancholy experience of men makes this warning a platitude. No elaborate community of elaborate organisms could exist unless its systems of symbolism were in general successful. (S 87)

I have developed the implications of this quote in the context of cultural symbolization elsewhere.[57] Let it only be said here that the *arbitrariness* of symbolisation that expresses mutual transcendence in the symbolic process will for evolution have the function of a necessary disconnection by which symbolism becomes a matter of *practice*, of empirical tests, and pragmatic satisfaction of truth (PR 181) by *survival*, not of the fittest, but of the most *sensitive*.[58] Here symbolism is, to use a theological term, *prophetic*; and this prophetic ability is based on perceptive sensibility – something that *dharmic* religious traditions value most highly as an expression of enlightenment and liberation. Ecological consciousness is such an ability to use arbitrariness as an artistic instrument in the development of sensitivity and, conversely, demonstrate this sensitivity by an artistic ability to create social organisations that include their environment in a most sensitive way so as not only to survive, but to establish that which Whitehead calls a civilisation of 'Truth, Beauty, Adventures, Art, Peace' (AI 285).

But the 'great societies of the universe' need not be of human nature, and if they are human, they might not show the evolutionary character referred to here. In any case, their evolution through novelty or their revolutionary character due to the continuing or disruptive forces at play leads Whitehead to a symbolisation of the prophetic nature of symbolism with both the appearance of reason and gravitation.

15. The great societies of the universe

> Reason can be compared to the force of gravitation, the weakest of all natural forces, but in the end the creator of suns and of stellar systems: – those great societies of the Universe . . . [The] doctrine of the disruptive tendency due to novelties, even those involving a rise to finer levels, is illustrated by the effect of Christianity on the stability of the Roman Empire. It is also illustrated by the three revolutions which secured liberty and equality for the world – namely the English revolutionary period of the seventeenth century, the American Revolution, and the French Revolution. (S 69–70)

As both forces, reason and gravitation, are an expression of creativeness, but in a weak way, that 'dwells upon the tender elements in the world, which slowly and in quietness operate by love' (PR 343), one is reminded of Walter Benjamin's weak messianic force or Derrida's permanently deferred justice.[59] They seem to induce a feeling of awe through their symbolic negotiations as a 'survival' in which outside and inside, multiplicity and unity, matter and form, materiality and consciousness, individual and society, organisation and life, organism and environment, the nested structure of a universe of interrelatedness and independence, mutual immanence and transcendence produce greatness through gentleness – a major concern of Whitehead in later years.[60] Reason, the 'art of life' of Whitehead's other small book *The Function of Reason* (FR 4), with *Symbolism* somehow surrounding *Process and Reality*, is envisioned with the same noble character of gravitation, a gentle balance between intensity and harmony – as is the case with the function of symbolic negotiation in such great societies. While suns and stars might be examples of this greatness achievable in the universe, reason only seems to give us hope that, if employed in the process of symbolic negotiation (but maybe not in the form of rationalism), it might spur on human society to similar nobility.[61]

In the meantime, the suspension of symbolism, although it makes space for more chaotic forms of novelty, also devastates the sensitivity necessary for cultural refinement. This is the ambivalence of the three revolutions Whitehead mentions in *Symbolism* – and here we also rediscover Gibbon's view on the impact of Christianity on the Roman Empire. The breakdown of symbolic repetition, the building of a character, so to say, culminating in an irruption, a breakdown, and a complete reconfiguration of symbolic meanings – arbitrary as the ones before, but more suited to the environments in which a new organisation wants to survive and, at the same time, sustain its environment – is the *one prophetic truth* of sociological nature. With its entreaty to develop human society consonantly with the great societies of the universe, Whitehead leaves us at the end of the book. This, in a sense, is Whitehead's symbolic testament.

16. An arrow in the hand of a child

> It is the first step in sociological wisdom, to recognize that the major advances in civilization are processes which all but wreck the societies in which they occur: – like unto an arrow in the hand of a child. The art of free society consists first in the maintenance of the symbolic code; and secondly in fearlessness of revision, to secure that the code serves those purposes which satisfy an enlightened reason. Those societies

which cannot combine reverence to their symbols with freedom of revision, must ultimately decay either from anarchy, or from the slow atrophy of a life stifled by useless shadows. (S 88)

May we find our way!

Notes

1. Faber, *God as Poet of the World*, §41.
2. Keller, *The Face of the Deep*, ch. 11.
3. Deleuze and Guattari, *What Is Philosophy?*, ch. 2; Faber, 'Whitehead at infinite speed', 39–72.
4. Faber, Henning and Combs, *Beyond Metaphysics?*
5. See on event, nexus and society, e.g., Kraus, *The Metaphysics of Experience*, 65–75.
6. Butler, *Giving an Account of Oneself*, 37.
7. Butler, 'On this occasion', 9.
8. Butler, *Gender Trouble*, 33.
9. See the conference 'Whitehead's Account of the Sixth Day' at http://www.youtube.com/watch?v=H5GAgZ1zIhc (last accessed March 2017).
10. Butler, *Gender Trouble*, 15; Faber, 'Introduction: negotiating becoming', 14.
11. Butler, *Gender Trouble*, 33.
12. Butler, 'On this occasion . . . ', 9.
13. Deleuze and Parnet, *Dialogues*, 57; Faber, *The Divine Manifold*, 78–84.
14. Faber, *The Divine Manifold*, 460–3; Faber, 'Khora and violence', 105–26.
15. Kraus, *The Metaphysics of Experience*, 76–7 and passim.
16. Butler, *Gender Trouble*, 30–3.
17. Deleuze, *Desert Islands and Other Texts 1953–1974*, 117–27.
18. Lango, *Whitehead's Ontology*, 18–46.
19. Adorno, *Negative Dialectics*, 149.
20. On the subjecting stabilisation of regulatory mechanisms by means of abstraction, see Butler, *Gender Trouble*, ch. 5.
21. Pailin, 'The meaning and use of abstraction in Whitehead', 81–8.
22. Gutting, *French Philosophy in the Twentieth Century*, 293–6.
23. Faber, 'Introduction: negotiating becoming'.
24. Although such a dissolution seems to contradict Whitehead's contention that actual entities and eternal objects are mutually exclusive in his scheme (PR 22), it is prepared by the identification of creativity as the form beyond all forms (PR 20) and the unifying term 'potentiality' in which real potentiality – that is, potentiality in act – and pure potentiality or possibility are not merely linguistically connected (PR 22–6). Similarly, Whitehead (1) undermines the dichotomy of universals and particulars by dislodging them from the categories of abstract and concrete, and

(2) reverses the classical association between activity on the one hand, and passivity and form and matter on the other. See Faber, *God as Poet of the World*, 61 and 171–2, as well as §§9–16 passim.
25 Rorty, 'Matter and event'.
26 Faber, 'Multiplicity and mysticism'.
27 Kraus, *The Metaphysics of Experience*, 75–93.
28 Faber, 'Immanence and incompleteness'.
29 Faber, '"Amid a Democracy of fellow Creatures"'.
30 Faber, 'Khora and violence'.
31 Butler, *Bodies That Matter*, ch.1; Butler, *Gender Trouble*, ch. 5.
32 Faber, 'Introduction: negotiating becoming', 40–3.
33 Keller, 'Introduction: the process of difference, the difference of process', 15–16.
34 Butler, *Gender Trouble*, ch. 5.
35 Faber, *The Divine Manifold*, chs 6, 8 and 10.
36 Faber, 'Tears of God'.
37 Deleuze and Guattari, *What Is Philosophy?*, 35.
38 This is not only resonant with *khora*, but with the axiom of Whitehead's philosophy that there is no relation beyond relation/relationality (PR 4); see Faber, 'Immanence and incompleteness', 102.
39 Faber, *The Divine Manifold*, ch. 13.
40 Faber, '"Amid a Democracy of fellow Creatures"'.
41 'Error' may relate not only to the more ontological statement in *Process and Reality* on the additional value of truth for a process (PR 259), if it adds interest and which otherwise heads for novelty by non-conformity to the past; rather, it may also relate to the fundamental importance in *Adventures of Ideas* of disharmony and discord for the whole process (AI 257); see Faber, 'O bitches of impossibility!'
42 Faber, *The Divine Manifold*, 213; Faber, *God as Poet of the World*, §11; see also Vattimo, *Beyond Interpretation*, ch. 1.
43 Faber, '"Amid a Democracy of fellow Creatures"', 203, 209, 216.
44 Gutting, *French Philosophy*, 298–304.
45 Kristeva, *Revolution in Poetic Language*, ch. 1.
46 Rinpoche, *Luminous Mind*, 151–63.
47 Inada, 'The metaphysics of Buddhist experience and the Whiteheadian encounter'; Inada, 'Whitehead's "actual entity" and the Buddha's anatman'.
48 Odin, *Process Metaphysics and Hua-Yen Buddhism*.
49 Faber, *The Divine Manifold*, 275.
50 Faber, *The Divine Manifold*, 337–8.
51 Faber 'Introduction: negotiating becoming', 36–43.
52 Faber, *The Divine Manifold*, ch. 9.
53 Faber, *Prozeßtheologie*, § 21.
54 Faber, *The Divine Manifold*, ch. 2 and ch. 13.

55 Faber, *The Divine Manifold*, ch. 6.
56 Faber, 'Ecotheology, ecoprocess, and ecotheosis'.
57 Faber, 'Cultural symbolization of a sustainable future'.
58 This recalls Derrida's last interview, in which he meditates on the notion of 'survival' in such a way that 'survival as a complication of the opposition death-life proceeds with me from an unconditional affirmation of life. Survival is life beyond life, life more than life, and the discourse I undertake is not death-oriented, just the opposite, it is the affirmation of someone living who prefers living, and therefore survival, to death; because survival is not simply what remains, it is the most intense life possible.' Derrida, *Learning to Live Finally*, 50–1.
59 Faber, 'Messianische Zeit'.
60 Faber, 'Three hundred years of Whitehead: halfway'.
61 Faber, 'Surrationality and chaosmos'.

Bibliography

Adorno, Theodor W., *Negative Dialectics* (London: Routledge, 1990).
Butler, Judith, *Bodies That Matter: On the Discursive Limit of 'Sex'* (New York: Routledge, 1993).
Butler, Judith, *Gender Trouble: Feminism and the Subversion of Identity* (New York: Routledge, 1999).
Butler, Judith, *Giving an Account of Oneself* (New York: Fordham University Press, 2005).
Butler, Judith, 'On this occasion . . .', in Roland Faber, Michael Halewood, and Deena M. Lin (eds), *Butler on Whitehead: On the Occasion* (Lanham: Lexington, 2012), pp. 3–17.
Cloots, Andre, and Keith A. Robinson (eds), *Deleuze, Whitehead and the Transformation of Metaphysics* (Brussels: Flemish Academy of Sciences, 2005).
Deleuze, Gilles, *Desert Islands and Other Texts 1953-1974*, ed. David Lapoujade (Los Angeles: semiotext(e), 2004).
Deleuze, Gilles, and Felix Guattari, *What Is Philosophy?* (New York: Columbia University Press, 1994).
Deleuze, Gilles, and Claire Parnet, *Dialogues* (New York: Columbia University Press, 1977).
Derrida, Jacques, *Learning to Live Finally: The Last Interview* (Brooklyn: Melville House Publishing, 2007).
Faber, Roland, '"Amid a Democracy of fellow Creatures" – onto/politics and the problem of slavery in Whitehead and Deleuze (with an intervention of Badiou)', in Roland Faber, Henry Krips and Daniel Pettus (eds), *Event and Decision: Ontology and Politics in Badiou, Deleuze, and Whitehead* (Cambridge: Cambridge Scholars Publishing, 2010), pp. 192–237.

Faber, Roland, 'Cultural symbolization of a sustainable future', in Adrian Parr and Michael Zaretsky (eds), *New Directions in Sustainable Design* (London, Routledge, 2011), pp. 242–55.

Faber, Roland, 'Ecotheology, ecoprocess, and *ecotheosis*: a theopoetical intervention', *Salzburger Zeitschrift für Theologie*, 12 (2008), 75–115.

Faber, Roland, *God as Poet of the World: Exploring Process Theologies* (Louisville: WJK, 2008).

Faber, Roland, 'Immanence and incompleteness: Whitehead's late metaphysics', in Roland Faber, Brian G. Henning and Clinton Combs (eds), *Beyond Metaphysics? Explorations in Alfred North Whitehead's Late Thought* (Amsterdam: Rodopi, 2010), pp. 91–107.

Faber, Roland, 'Introduction: negotiating becoming', in Roland Faber and Andrea M. Stephenson (eds), *Secrets of Becoming: Negotiating Whitehead, Deleuze, and Butler* (New York: Fordham University Press, 2011), pp. 1–49.

Faber, Roland, 'Khora and violence: revisiting Butler with Whitehead', in Roland Faber, Michael Halewood and Deena M. Lin (eds), *Butler on Whitehead: On the Occasion* (Lanham: Lexington, 2012), pp. 105–26.

Faber, Roland, 'Messianische Zeit. Walter Benjamins "mystische Geschichtsauffassung" in zeittheologischer Perspektive', *MThZ*, 54 (2003), 68–78.

Faber, Roland, 'Multiplicity and mysticism: toward a new mystagogy of becoming', in Nicholas Gaskill and A. J. Nocek (eds), *The Lure of Whitehead* (Minneapolis: University of Minnesota Press, 2014), pp. 187–206.

Faber, Roland, '"O bitches of impossibility!" – Programmatic dysfunction in the chaosmos of Deleuze and Whitehead', in K. Robinson (ed.), *Deleuze, Whitehead, Bergson: Rhizomatic Connections* (London: Palgrave Macmillan, 2009), pp. 200–19.

Faber, Roland, *Prozeßtheologie. Zu ihrer Würdigung und kritischen Erneuerung* (Mainz: Gruenewald, 2000).

Faber, Roland, 'Surrationality and chaosmos: for a more Deleuzian Whitehead (with a Butlerian intervention)', in Roland Faber and Andrea M. Stephenson (eds), *Secrets of Becoming: Negotiating Whitehead, Deleuze, and Butler* (New York: Fordham University Press, 2011), pp. 157–77.

Faber, Roland, 'Tears of God – in the rain with D. Z. Philips and J. Keller, waiting for Wittgenstein and Whitehead', in Randy Ramal (ed.), *Metaphysics, Analysis, and the Grammar of God* (Tübingen: Mohr Siebeck, 2010), pp. 57–103.

Faber, Roland, *The Divine Manifold* (Lanham: Lexington, 2014).

Faber, Roland, 'Three hundred years of Whitehead: halfway', *Process Studies*, 41:1 (2012), 5–20.

Faber, Roland, 'Whitehead at infinite speed: deconstructing system as event', in Christine Helmer, Marjorie Suchocki and John Quiring (eds), *Schleiermacher and Whitehead: Open Systems in Dialogue* (Berlin: DeGruiter, 2004), pp. 39–72.

Faber, Roland, Brian G. Henning and Clinton Combs (eds), *Beyond Metaphysics? Explorations in Alfred North Whitehead's Late Thought* (Amsterdam: Rodopi, 2010).
Gutting, Gary, *French Philosophy in the Twentieth Century* (Cambridge: Cambridge University Press, 2001).
Inada, Kenneth, 'The metaphysics of Buddhist experience and the Whiteheadian encounter', *Philosophy East and West*, 25 (1975), 465–88.
Inada, Kenneth, 'Whitehead's "actual entity" and the Buddha's anatman', *Philosophy East and West*, 21 (1971), 303–16.
Keller, Catherine, 'Introduction: the process of difference, the difference of process', in Catherine Keller and Anne Daniell (eds), *Process and Difference: Between Cosmological and Poststructuralist Postmodernism* (New York: State University of New York Press, 2002), pp. 1–30.
Keller, Catherine, *The Face of the Deep: A Theology of Becoming* (New York: Routledge, 2013).
Kraus, Elizabeth, *The Metaphysics of Experience: A Companion to Whitehead's Process and Reality* (New York: Fordham, 1998).
Kristeva, Julia, *Revolution in Poetic Language* (New York: Columbia University Press, 1984).
Lango, John W., *Whitehead's Ontology* (Albany: State University of New York Press, 1972).
Odin, Steve, *Process Metaphysics and Hua-Yen Buddhism: A Critical Study of Cumulative Penetration vs. Interpenetration* (Albany: State University of New York Press, 1982).
Pailin, Isabella, 'The meaning and use of abstraction in Whitehead', in Andre Cloots and Keith A. Robinson (eds), *Deleuze, Whitehead and the Transformation of Metaphysics* (Brussels: Flemish Academy of Sciences, 2005), pp. 81–8.
Rinpoche, Kalu, *Luminous Mind: The Way of the Buddha* (Boston: Wisdom Publications, 1997).
Robinson, Keith A. (ed.), *Deleuze, Whitehead, Bergson: Rhizomatic Connections* (Chippenham: Palgrave Macmillan, 2009).
Rorty, Richard, 'Matter and event', in Lewis S. Ford and George L. Kline (eds), *Explorations in Whitehead's Philosophy* (New York: Fordham University Press, 1983), pp. 68–103.
Vattimo, Gianni, *Beyond Interpretation: The Meaning of Hermeneutics for Philosophy* (Stanford: Stanford University Press, 1997).
Whitehead, Alfred North, *Adventures of Ideas* (New York: The Free Press, [1933] 1967).
Whitehead, Alfred North, *Process and Reality: An Essay in Cosmology*, corrected edition, ed. David Ray Griffin and Donald W. Sherburne (New York: The Free Press, [1929] 1978).
Whitehead, Alfred North, *Religion in the Making* (New York: Fordham University Press, [1926] 1996).

Whitehead, Alfred North, *Science and the Modern World* (New York: The Free Press, [1925] 1967).
Whitehead, Alfred North, *Symbolism: Its Meaning and Effect* (New York: Fordham University Press, [1927] 1985).
Whitehead, Alfred North, *The Principle of Relativity: With Applications to Physical Science* (Cambridge: Cambridge University Press, [1922] 2011).

Part II
Adventures in Culture and Value

4

The Inhumanity of Symbolism

MICHAEL HALEWOOD

One aim of this chapter is to protect Whitehead's thought from any charge of humanism. Throughout his work, Whitehead tries to avoid any easy dualisms which might lead us to venerate the special aspects or role of humans in the world, at the expense of giving due weight to the role of 'stubborn fact' (PR xiv and passim). Advocating a simple version of humanism would mean that nature has 'bifurcated' (CN 26–48), that humans and their unique properties are different and distinct from all other elements of the world. This is not to suggest that Whitehead was an anti-humanist, as this would also run the risk of allowing nature to bifurcate. Emphasis would be laid on the side of brute matter of fact, with the associated difficulty of establishing how meaning can be put back into a mute world which has no value of its own.

The more specific concern, which runs throughout this chapter, is that a focus on Whitehead's text *Symbolism: Its Meaning and Effect* runs the risk of inadvertently creating a sophisticated version of humanism. Such a humanism might not start with the human, might not even make the human central to our understandings of meaning, but it could let a version of human exceptionalism slip in through the back door – for example, if too much emphasis is placed on how language indicates the way in which humans express themselves symbolically. This is not to deny any link between humans, symbolism and language. Whitehead makes it quite clear that there is such a link. My point is that we need to be careful when approaching the relations between humans and symbolism. The main point of interest is the inhumanity involved in making us human. As Stengers puts it, it is a question of 'what we became when we were given speech, not what was given to us by speech'.[1]

Whitehead on Symbolism

Whitehead is very careful with his use of words, and it is worth paying attention to even slight variations in his terms. One notable example is on page 1 of *Symbolism*, where he states that during the Middle Ages

> symbolism *seemed* to dominate men's [sic] imaginations. Architecture was *symbolical*, ceremonial was *symbolical*, heraldry was *symbolical* . . . But such symbolism is on the fringe of life. (S 1, emphasis added)

It is important to note that Whitehead makes a sharp but subtle distinction between symbolism and the symbolical. In the Middle Ages, symbolism *seemed* to dominate human imaginations, and the evidence for this still appears to exist in the old buildings, flags and insignias which pepper Europe in its cathedrals, parliaments and universities. The mistake to avoid is taking these too literally as examples of symbolism. They were, and are, elements of the kind of symbolism which interests Whitehead, but they are not its core; they are on the fringes of life. The roots and power of symbolism lie elsewhere. Whitehead continues: 'There are deeper types of symbolism, in a sense artificial, and yet such that we could not get on without them' (S 2). Importantly, these deeper forms of symbolism are 'artificial' (in a sense). They do not spring immediately from 'nature'. Yet, us humans require them, as 'we could not get on without them'. The examples that Whitehead gives are 'language and algebra [which] seem to exemplify more fundamental types of symbolism than do the Cathedrals of Medieval Europe' (S 2). This is not the end of the argument. Whitehead moves beyond these artificial forms to the *one* form of symbolism which is even more fundamental. This is the symbolism involved in perception. There is one important comment to make about this.

Perception is a part of the world. It is something in which humans partake, and this needs to be explained. But human perception is not originary. Symbolism is more widespread than that. It can be found in dogs and tulips (S 4–5). The reason for this is that, according to Whitehead, symbolism is tied up with what he calls 'sense-presentation', and this is something which is not to be found in humans alone. If a dog sees a coloured shape it would, or could, react to it, for example by jumping on it (the coloured shape is taken to be a place to rest, what we would call a chair). 'Also a tulip which turns to the light has probably the very minimum of sense-presentation' (S 5). All in all, Whitehead's point is that 'Symbolism from sense-perception to physical bodies is the most natural and widespread of all symbolic modes' (S 4). But it should be noted that the symbolism involved in such sense-perception is not the only kind of symbolism. There are others. It is time to produce a definition of symbolism.

Defining Symbolism

When Whitehead does come to define symbolism, he starts by talking of the human mind (S 7–8). While Whitehead's aim at this point in his argument might well be to focus on the relation between symbolism and human life, this does not mean that symbolism can be understood only in terms of human life. Quite the opposite. In order to understand symbolism in human life, we have to locate symbolism in its wider context. Maintaining this distinction is not always easy, as Whitehead strays from discussions of symbolism as a wide element of existence to a closer treatment of the operations of symbolism within the human realm. The signposts that Whitehead provides are not always as clear as they could be, and Whitehead often switches between his more general account and the human elements of symbolism. This should be borne in mind during the following discussions.

Symbols are tied up with meaning. The main point of a symbol is that it means something. But what constitutes meaning? Meaning is that which is elicited by a symbol, be it in terms of usage, belief or consciousness. The key word here is 'elicited'. A symbol is something which provokes a response; it produces another experience. In short, the operation of symbolism is the move from one element of experience to another. The first element is the symbol, the second element is the meaning of the symbol. This is a very democratic account. Anything can be a symbol, anything can be a meaning. If we hear the word 'tree' we might think of a tree. In this case, the word 'tree' is a symbol and our thought of a tree is its meaning. Or, we could be working for a tree-felling company, where our role is to chop down trees. On seeing the word 'tree' on a placard in front of an object, we might move straight from seeing this to starting up our chainsaw. Conscious thought does not have to intervene. One element of experience (reading the word 'tree') has elicited another, the chopping down of the tree. Here, the meaning of the symbol 'tree' is the chopping down of the tree. This is what I take Whitehead to mean when he says: 'There are no components of experience which are only symbols or only meanings' (S 10).

> Further, why do we say that the word 'tree' – spoken or written – is a symbol to us for trees? Both the word itself and trees enter into our experience on equal terms; and it would be just as sensible, viewing the question abstractedly, for trees to symbolize the word 'tree' as for the word to symbolize the trees. (S 11–12)

It is important, however, to be precise. It might seem that I am suggesting that symbolism is merely a case of one experience being followed by another. It is not this simple (or uninteresting). The two components, the symbol and the meaning, actually comprise one experience. They are 'two components in a complex experience' (S 10). Whitehead has moved on in his argument (though he does not tell us that he has). He is now talking about 'symbolic reference'.

Whitehead defines symbolic reference as follows: 'The organic functioning whereby there is transition from the symbol to the meaning will be called "symbolic reference"' (S 8). It is unusual for Whitehead to provide definitions, and the fact that he does so here should alert us to the importance of the term. Whitehead continues: 'This symbolic reference is the active synthetic element contributed by the nature of the percipient' (S 8). Note that despite Whitehead's mention of the human mind in the previous paragraph, he is now talking only of a 'percipient' and, as he has already made clear, such percipients are not limited to humans – they include dogs and tulips, at least. The more general aspect of Whitehead's argument is made more explicit a couple of sentences later when he writes: 'We must conceive perception in the light of a primary phase in the self-production of an occasion of actual existence' (S 8). Those who have read *Process and Reality* will feel themselves to be on more familiar ground. And it is this ground that I want to emphasise. Let us forget humans for the moment and take Whitehead at his word, but metaphysically.

For symbolic reference to operate, there must be something in common between the symbol and its meaning. More than this, this 'something held in common' produces a real ground. Symbolic reference 'requires a ground founded on some community between the natures of the symbol and the meaning' (S 8). This is only the first step. This common ground does not mean that there will be, or has to be, symbolic reference. For symbolic reference to occur, the two elements – the symbol and the meaning – must be incorporated in one act of experience. Active self-production through the combining of elements which were diverse into a temporary unity is, for Whitehead, the hallmark of all existence, not just of symbolism. Symbolic reference, therefore, is just one example of a more general aspect of reality.

Humans will obviously be involved in this, but they are not alone in such situations. Human symbolism is not a special or exceptional case. This 'is the foundation of a thorough-going realism. It does away with any mysterious element in our experience which is merely meant' (S 10). Symbols, symbolism and meaning are not mysterious or magical in themselves, though they might elicit experiences which we consider to be magical, such as love, trust or a belief in fairies. Yet such symbolism is based in reality. When Whitehead does return to a direct discussion of human experience, he introduces the word 'objective' to reinforce this point (S 13).

Before proceeding, I want to touch on a problem that some encounter when interpreting Whitehead. With his emphasis on the unity of an act of experience, the irreducibly atomic individuality of his actual entities, it is possible to read Whitehead as prioritising the individual at the expense of the social. Of course, we have the testimony of Whitehead's extensive comments on the social environment as inherent in the creation of actual entities to counter such claims (see, for example, PR 203-5). Nevertheless, there is, for some, an apparent problem in reconciling the utter individuality of occasions of experience with what we might more commonly think of as social existence (economies, education, unemployment, etc.). In one sense, Whitehead does try to discuss such human social phenomena in Chapter III of *Symbolism*, where he talks of public ceremonies, national life, human societies and revolution. These passages, however, do not respond to the problem that I am trying to focus on in this chapter, that is, the inhumanity of the social. In order to provide a fuller account of Whitehead's position, I will now take a detour through some of the thoughts of Roland Barthes, Michel Foucault and Raymond Williams. Hopefully, this will enable me to return to Whitehead's text in the conclusion in a position to make a stronger argument.

The Inhumanity of the Social

For many sociologists or social theorists, talking of symbolism, symbols and meanings will lead us inexorably to Roland Barthes. In his important work *Mythologies*, Barthes sets out to unravel and display the complex ways in which the contemporary world is suffused by signs. In doing so, he builds on the work of Ferdinand de Saussure (1857–1913). It is signs that we confront in the everyday world. The operation of signs, or 'signification', is not an innocent affair. It is certainly not a simple question of identifying what a sign points to or indicates in the world, as for Barthes there is no such thing as pure denotation. Nevertheless, signs are involved in processes of signification through which meaning is generated. These processes are located within specific social and historical situations and forces. As a result, there is no direct access to an objective reality. There is no pure way of seeing or understanding the world. Consequently, for Barthes, the connotations of a sign are more important than its denotation. Connotations are determined by the codes to which the interpreter has access. Therefore, meaning and our understanding of meaning are always encoded within signs, as meaning is always part of a codified system. The cultural codes which we find ourselves in are not innocent; they are tied up with the interests of specific groups, with systems of understanding that are related to systems of power. More specifically, for Barthes cultural codes are ideological. Semiology (the study of signs) can uncover the ideological dimension of signs in popular culture.[2]

Throughout the essays in his book, Barthes takes apparently mundane contemporary phenomena such as margarine or wrestling and interrogates how they operate as part of a wider system of signification. His calls these phenomena 'myths', and he uses this term in a very specific sense. It is by considering the contemporary world as mythical that we can begin to unpack how specific social and historical elements are passed off as natural. 'We reach here the very principle of myth: it transforms history into nature'.[3] It is because symbolism is so widespread that it goes unnoticed. Barthes asks us, as does Whitehead, to pay more attention to the operations of symbols and meanings in the world. Unlike some traditional conceptions of myths, for Barthes they are not ambiguous: 'Myth does not deny things, on the contrary, its function is to talk about them'.[4]

It is here that the inhumanity of symbolism returns, though now it has become a matter of the inhumanity of the social. What does this mean? The signs and symbols that we encounter in the contemporary world are not of our own immediate making. We make sense of the world, we understand it, through signs which themselves have gone through a process. It is a process which organises the world so that it makes sense to us and enables us to make sense of ourselves in the world. As with Whitehead, it is not that we are simply human; we become human. For Barthes, this involves a separation of ourselves from the processes inherent in the making of our world so that the world appears to be made up of things which mean something separately from us. There is a double inhumanity here. One element is that there are other things involved in the process by which we become contemporary humans. This element is not so problematic; after all, it is one founding element of Whitehead's philosophy. The other, more insidious aspect of this inhumanity is that in becoming human we separate ourselves from the things of the world. For Barthes, this process involves the passing from history to nature. As a result, we believe that we inhabit a world which is innocent, pure, without contradiction. We think that things mean things on their own terms. In doing so, we downgrade the human element, the specific historical and cultural forms of the things of the world. We deny the humanity involved in the world and, in doing so, we limit our understanding of our own humanity. Thinking of the world as full of objects which are separate from us is an example of what Whitehead calls 'the Bifurcation of Nature' (CN 26–48). Our tacit acceptance of this worldview limits our thoughts and our actions. This limitation is a form of inhumanity.

This argument could be seen as another version of Marx's concept of the fetishism of the commodity.[5] We take commodities to be self-sufficient entities, to be simple objects, and fail to see the human (or social) relations that have gone into making them. When we buy a commodity, we do not see the factory systems, the hours worked, the conditions of the workers that went into making the things that we take from the shelves. We see things that we can buy and own, failing to realise that the possibility to purchase and possess is not natural: it is only one way of organising the world. The problem, for me at least, with Barthes' argument is not that he retains an important element of Marxism within his texts. Rather, it is that Barthes relies upon a sophisticated but stark use of the notion of ideology. The manner in which signs are imposed upon us indicates, for Barthes, the operations of ideology. 'Bourgeois ideology . . . turns culture into nature', he declares.[6] This firm reliance upon the concept of ideology is problematic; it is a concept with a long and tortuous history. Instead of revisiting this history, I will attempt to come at the issue from a different angle. It is not a matter of reclaiming or remodelling the concept of ideology. My aim is to avoid the term ideology while recognising the problems which led to ideology being posited as a solution. We need to recognise what we first wanted from the concept and focus on the remnants of the problem which led to its development. More importantly, we need to establish how these elements can be rethought. One way of doing this is through the writings of Foucault.

Foucault on Ideology

> I don't think that we should consider the 'modern state' as an entity which was developed above individuals, ignoring what they are and their very existence, but . . . as a very sophisticated structure, in which individuals can be integrated . . . and submitted to a set of very specific patterns.[7]

This may not be the most incisive of Foucault's statements, but it does tell us something about the context in which he was writing. Those Marxists (especially the Maoists and members of the French Communist Party who were so influential in French academia in the late 1960s and early 1970s[8]) who saw the 'modern state' as the source of all our problems, and the site of a central production of ideology, missed their target. Worse, in doing so, they missed the experiences, the real lives, of those people in whose name they built their critique. Foucault's aim is to redress the balance, to shift focus, to produce a more accurate account. There are, he says, patterns of experience, rather than any simplistic model of a structural base and ideological superstructure. I will return to this notion of 'patterns of experience' in the concluding section of this chapter.

Foucault did not think that he was betraying Marx. He made this clear in an interview given shortly after the publication of his book *Discipline and Punish* (1975) when he said: 'I often quote concepts, texts and phrases from Marx without feeling obliged to add the authenticating label of a footnote with a laudatory phrase'.[9] Nevertheless, Foucault was wary of any unthinking use of the concept of ideology: 'basically I do not believe that what has taken place can be said to be ideological. It is both much more and much less than ideology'.[10]

Ideology exists. We should take it seriously, but not as the be-all-and-end-all. In itself ideology does not explain what is going on, what constitutes the contemporary world or our place in it. To return to Whitehead for a moment, we must not commit the sin of 'misplaced concreteness' by taking ideology to be a real thing when it is in fact an abstraction (SMW 64). Ideology might indicate that there is a problem, but it is not the problem in itself. We need to shift focus, but not lose sight of what drew us to posit ideology as the solution. Foucault attempts to do this through his reconceptualisation of power, which has been both important and troublesome for sociology and social theory ever since. This is not the path that I want to follow. Instead, I will make another jump, one which will eventually bring us back to Whitehead, after a detour through Raymond Williams' notion of 'structures of feeling'.

Raymond Williams and Structures of Feeling

I have just tried to argue that we need to re-approach that which led us to think that ideology was a solution. In this section, I will try to show how, as Foucault puts it, what concerns us 'is both much more and much less than ideology'. This will require some conceptual work, but does not involve a wholesale rejection of Marx or Marxism. This is one reason why I have chosen to discuss the inveterate but subtle Marxist Raymond Williams.

Raymond Williams (1921–88) was an important Marxist cultural theorist and historian. In his book *Marxism and Literature* (ML), Williams talks of what he calls 'Structures of Feeling'. His first move is to suggest that this notion is designed to avoid the tendency in cultural and social analyses to talk of culture in the past tense. Always linking culture to the past, to tradition, involves the 'conversion of experience into finished products' (ML 128). This also means that we tend to treat both culture and society as complete, as over, as objects. According to Williams, this misses a vital aspect of culture, namely, that it is active and ongoing. We look for 'formed wholes rather than forming and formative processes' (ML 128). The task he sets himself is to

> find other terms for the undeniable experience of the present: not only the temporal present, the realization of this and this instant, but the specificity of present being, the inalienably physical, within which we may indeed discern and acknowledge institutions, formations, positions, but not always as fixed products, defining products. (ML 128)

There is a need to account for how we are what we are now, and what we are becoming (this might be one way of summarising the work of Foucault). We need to be able to describe process. This does not mean falling into relativity or any notion of continuous becoming. Within the present we can still discern wider aspects which suffuse our present experience, and elements of this are 'inalienably physical'. The term that Williams chooses to describe these processes is 'Structures of Feeling'. This is a term that resonates with elements of Whitehead's work. Nevertheless, there is also a useful difference between the work of these two thinkers.

Williams makes it explicit that he wants to account for what I have put under the banner of 'ideology'. But he does so in a manner which is less rigid than that of Barthes: '"feeling" is chosen to emphasize a distinction from more formal concepts of "world-view" or "ideology"' (ML 132). Through his use of the phrase 'structures of feeling', Williams is trying to get at 'a kind of feeling and thinking which is indeed social and material, but each in an embryonic phase before it can become fully articulate and defined exchange' (ML 131). He has not presupposed what is social about such feeling, and he has not made it reliant solely upon human consciousness or imagination. The feelings involved are material, yet they are coming to be. Their active status, rather than their settled character, is what makes them so important and informative.

Such feelings can be identified by 'a particular quality of social experience and relationship, historically distinct from other particular qualities' (ML 131). The emphasis upon quality is crucial. It is quality which grants specificity to a feeling and enables us to identify it. This quality of a feeling is different from its content. It is a way of feeling and points to the importance of the adverbial. *How* things happen is as important as *what* things happen. This 'how' is adverbial.[11] Williams keeps this argument within a Marxist remit. There is a quality to the feelings of capitalism which make it distinct from the feelings of other epochs. Under capitalism, there is a tendency for our feelings to manifest a specific quality. This is not a form of economic reductionism, I would argue. Feelings are not ideological productions which are somehow produced by the dominant class (or 'the 1%'), forcing the rest of us (the so-called '99%') to think and feel in a specific way, a way that does not benefit us but keeps us in our place. This is where it is important to return to Whitehead's approach and to use him to defend Williams from the charge of advocating a traditional or reductionist Marxist position. Feelings are real; they are an element of the process by which we become what we are. Feelings make us what we are by incorporating elements of other things, of what we are not, into what we are now. This is, again, an aspect of the 'inhumanity of symbolism'. What is specific about symbolism is that it puts us in the realm of meaning: 'we are concerned with meanings and values as they are actively lived and felt' (ML 132). As Whitehead has already told us, the following of one experience by another makes the former the symbol and the latter the meaning. Such meaning is located not just in our minds but in our bodies, our actions.

In order to explain his anti-reductionist Marxism, Williams argues that the different qualities of feeling that we find in history should *not* be treated as 'epiphenomena of changed institutions, formations . . .

they are from the beginning taken as *social* experience, rather than as "personal" experience or as the merely superficial or incidental "small change" of society' (ML 131). Williams' first move is to depersonalise feelings precisely in order to make them social. This depersonalisation is not part of the inhumanity of the social; it is an argument against reducing feelings to the merely 'subjective', and thus less real or important than the 'objective'. Feelings are real. Feelings are social. Feelings tell us something about the world and us. To analyse them we need a different methodology from that which is often used in social and cultural analyses, with their tendency to look only at the past. Feelings are not settled; they are not historical elements; they are ongoing, but they can still tell us something important.

The reader has likely noted that in the previous paragraph the word 'feelings' has been used as if it were interchangeable with 'experiences'. I believe that it is possible to make such a conflation, as Williams makes it clear that in place of 'structures of feeling', an 'alternative definition would be structures of *experience*' (ML 132; original emphasis here and below). The overlap between feeling and experience is productive when approaching this topic from a Whiteheadian perspective. There is also a link to Whitehead's discussion of 'conformation' as set out in *Symbolism*. Whitehead's approach fits nicely with Marx's assertion that 'Men [sic] make their own history, but they do not make it just as they please; they do not make it under circumstances chosen by themselves, but under circumstances directly encountered, given and transmitted from the past'.[12] There is a creativity involved in all that we do, but not an unfettered creativity. We need to distinguish between 'general' and 'real' potentiality (PR 65).[13] In *Symbolism*, conformation takes on the role of real potentiality. There is a need to pay attention to the part played by the past in the present; it provides the ground from which we and our world arise, but it is not a purely determining ground. It is in this vein that it is possible to read Williams' account of the changes in structures of feelings from one epoch to another, which he describes as '*changes of presence* [and . . .] although they are emergent or pre-emergent, they do not have to await definition, classification, or rationalization before they exert palpable pressures and set effective limits on experience and on action' (ML 132). In the dual reading of Whitehead and Williams that I am trying to set out, Whitehead's notion of 'conformation' becomes 'palpable pressures' which set the limits for our actions and the world.

Where Williams differs from Whitehead is in his insistence that we trace the specific qualities of feeling which exhibit the irruption of a certain way of thinking and feeling which is identifiable under capitalism. In doing so, we are looking for *structures* of feeling. 'We are then defining these elements as a "structure": as a set, with specific internal relations, at once interlocking and in tension' (ML 132). This mention of structure might seem problematic. *Post*-structuralism has warned of many of the pitfalls of a structuralist account. The concept of structure is especially troublesome for those Marxists who want to move beyond the notion of a base (structure) and superstructure. It might seem to condemn such Marxists to an admission that the economic base does comprise some kind of founding structure after all. This is why, although Whitehead does not rely on any notion of structure to make his arguments in *Symbolism*, it seems that he has something important to offer. In his calm and understated way, and through the use of a nice double negative, he reorients the problem:

> The components of experience are not a structureless collection indiscriminately brought together. Each component by its very nature stands in a certain potential scheme of relationships to the other components. It is the transformation of this potentiality into real unity which constitutes the actual concrete fact which is the experience. (S 86)

Components of experience and of feeling are 'not structureless', but nor are they 'structure-ful'. There is a move from potentiality into actuality, and into a unity of experience. This involves a transformation: transformation involves intensification, from which 'other elements of experience may arise' (S 86). Again, the production of other elements of experience is a key element of symbolism.

One important question remains: Whitehead has provided a neat philosophical account, so why have I compared his work with that of Raymond Williams? Is this just a matter of an interesting similarity between elements of their arguments? My answer to such questions would be that while I do find the similarities interesting, there is much more to it than that. Williams' key contribution can, perhaps, be found in the following passage:

> Yet we are also defining a social experience which is still *in process*, often indeed not yet recognized as social but taken to be private, idiosyncratic, and even isolating, but which in analysis . . . has its emergent connecting, and dominant characteristics, indeed its specific hierarchies. (ML 132)

Whitehead rarely uses the word 'hierarchy', though he does, intriguingly, mention 'a hierarchy of categories of feeling' (PR 166). Williams is much more interested in the role and status of hierarchies and in the 'dominant characteristics' that accompany them. This is an indication of the specific kind of Marxism that Williams is trying to develop, one where the feelings and experiences which are involved in such hierarchies are 'still *in process*'. The mention of hierarchies also relates to the concept of ideology. This is something that Williams recognises, but he still wants to keep some distance. His argument is not about 'casting off an ideology, or learning phrases about it, but confronting a hegemony in the fibres of the self and in the hard practical substance of effective and continuing relationships' (ML 212). As was seen in the discussion of Foucault, the concept of 'ideology' is distracting. It embroils people in trying either to throw it off or to indulge in complex, wordy arguments about it. This misses the main point, which is to focus on how the very fibres of the self, under capitalism, are produced and infected within relationships which are not of our own making, which include a level of dominance, and which involve a specific element of inhumanity in the humans that we are now.

Conclusion

By intertwining Whitehead and Williams, we can identify patterns of unevenness in existence and experience. There are even hierarchies. All these express the force of the problem which the concept of ideology has often been used to explain. However, using the work of Whitehead and Williams, we can rethink and re-approach this problem without having to get embroiled in the thorny question of ideology. Whitehead and Williams have identified the lumpiness inherent in the processes of existence. These take on a specific character under capitalism and are felt, bodily and mentally. 'Patterns of unevenness' is a phrase which not only avoids some of the problems associated with that of 'structure', it also chimes with the earlier quotation from Foucault.

Ultimately, I want to argue that there is an inhumanity involved in the process by which we become human. We need to focus on the patterns of lumpiness which constitute existence. For Whitehead, such unevenness is a necessary aspect of his metaphysical account of existence. For Williams, there is evidence of a specific lumpiness in our contemporary culture. To my mind, Williams enables us to make a more specific point than Whitehead, namely, that under capitalism a very peculiar aspect of inhumanity is involved in making us human. There is an unevenness to the symbols and meanings which help make us what we are; this unevenness is implicated in, and sustained by, the processes of capitalism. We require symbols to become what we are. Symbolism always involves an element of inhumanity. The specific operations of capitalism involve making one possible aspect of humanity an element within the inhumanity which makes us what we are today.

Notes

1 Stengers, 'Whitehead's account of the sixth day', 50.
2 Barthes, *Mythologies*.
3 Barthes, *Mythologies*, 129.
4 Barthes, *Mythologies*, 143.
5 Marx, *Capital: Volume I*, 163–77.
6 Barthes, *S/Z*, 206.
7 Foucault, 'The subject and power', 214.
8 Including Foucault's teacher, Louis Althusser.
9 Foucault, *Power/Knowledge*, 52.
10 Foucault, *Power/Knowledge*, 102.
11 See: Shaviro, *Without Criteria*, 38, 46; Halewood, *A. N. Whitehead and Social Theory*, 27–9; Halewood, *Rethinking the Social Through Durkheim, Marx, Weber and Whitehead*, 152–5.
12 Marx, *Selected Writings*, 300.
13 Halewood, *A. N. Whitehead and Social Theory*, 37–8.

Bibliography

Barthes, Roland, *Mythologies*, trans. Annette Lavers (London: Jonathan Cape, 1972).
Barthes, Roland, *S/Z* (Oxford: Basil Blackwell, 1974).
Foucault, Michel, *Power/Knowledge: Selected Interviews and Other Writings 1972–1977*, ed. Colin Gordon (Hemel Hempstead: Harvester Press, 1980).
Foucault, Michel, 'The subject and power', in Hubert Dreyfus and Paul Rabinow (eds), *Michel Foucault: Beyond Structuralism and Hermenuetics* (Chicago: University of Chicago Press, 1982), pp. 208–26.
Halewood, Michael, *A. N. Whitehead and Social Theory: Tracing a Culture of Thought* (London: Anthem Press, 2011).
Halewood, Michael, *Rethinking the Social Through Durkheim, Marx, Weber and Whitehead* (London: Anthem Press, 2014).
Marx, Karl, *Capital: Volume I* (London: Penguin Books, 1990).
Marx, Karl, *Selected Writings*, ed. David McLellan (Oxford: Oxford University Press, 1977).
Shaviro, Steven, *Without Criteria: Kant, Whitehead, Deleuze, and Aesthetics* (Cambridge, MA: MIT Press, 2009).

Stengers, Isabelle, 'Whitehead's account of the sixth day', *Configurations*, 13:1 (2005), 35–55.

Whitehead, Alfred North, *Process and Reality: An Essay in Cosmology*, corrected edition, ed. David Ray Griffin and Donald W. Sherburne (New York: The Free Press, [1929] 1978).

Whitehead, Alfred North, *Science and the Modern World* (Cambridge: Cambridge University Press, [1925] 1932).

Whitehead, Alfred North, *Symbolism: Its Meaning and Effect* (New York: Macmillan, 1927).

Whitehead, Alfred North, *The Concept of Nature* (Cambridge: Cambridge University Press, [1920] 1964).

Williams, Raymond, *Marxism and Literature* (Oxford: Oxford University Press, 1977).

5

Reverence, Revision and Creaturely Life: Whitehead's Political Theology of Enjoyment

BEATRICE MAROVICH

The figure of the creature, in the work of Alfred North Whitehead, functions as an almost superficial or superfluous category. The creature serves, primarily, as a cognate term for the actual entity or the 'individual fact' (PR 20):[1] the concretion left in the wake of creativity. It is as if the figure of the creature is merely a shadow, a reminder that actual entities are conditioned by creativity itself. God, like other actual entities, is a creature rather than a creator – God is transcended by creativity (PR 88).[2] And yet God still carries other ontological functions in Whitehead's philosophy. The figure of the creature, on the other hand, carries no ontological function that is all its own. It seems to float, like decorative adornment, throughout Whitehead's corpus.

The questions I pursue here, then, are quite rudimentary. What might the function of the creature have *been*, for Whitehead, in the first place? Does the creature play a conceptual role, beyond a shadow concept for creativity? Does it function as anything more than adornment? I argue that the creature is important in Whitehead's philosophy, perhaps above all *as symbol*. It is when we examine the creature *as symbol* (rather than, say, as ontological category) that its more integral features begin to emerge. Moreover, when the symbolic contours of creatureliness become visible in Whitehead's thought, what also becomes visible is the figure of a creature who symbolises resistance to a despotic deity. We see, instead, not a creature who cuts ties with the divine, but one who sources divinity for enjoyment. To the extent that Whitehead's own conceptual resistance against this form of divinity is informed by (and responsive to) Western intellectual history and politics, I have been tempted to call this a political theology.

Symbolism itself, like the figure of the creature in his work, 'has an unessential element in its constitution', Whitehead argued (S 1). It is almost arbitrarily discarded and adopted. And yet Whitehead also argued that when we reduce the process of symbolisation down to its bare bones, what we are left with are trains of reference (S 7). Symbolism, perhaps, is where we find the genetic strains and traces of cultural value. Symbolism itself is arbitrary. But symbols are culturally chosen and reinvented. Thus, symbolism reveals something about social life – especially its loyalties and disloyalties – when we examine these trains of reference.

Whitehead seemed wary of symbolisation on a number of levels. It has the tendency, he cautioned, to 'run wild'. This is why he found it socially necessary to engage in a 'constant pruning' of symbols. Indeed, he said, 'the successful adaption of old symbols to changes of social structure is the final mark of wisdom in sociological statesmanship' (S 61). And this pruning included, of course, a periodic revolution – cutting away all the branches of the shrub so that it might grow anew. 'Those societies which cannot combine reverence to their symbols with freedom of revision, must ultimately decay either from anarchy, or from the slow atrophy of a life stifled by useless shadows' (S 88). Whitehead's own combination of reverence and fearlessness of revision is very much on display in his symbolisation of the creature.

In what follows, I pursue my broader argument through two related threads. First, I suggest that not only does Whitehead's use of the creature expose a kind of reverence for theological tradition, but it also seems to expose his reverence for the literary tradition of British romantics – especially his reverence for their fearlessness of revision. Whitehead's creature is caught up in the trains of reference running through this literary tradition. Hence, the structure of creatureliness in Whitehead's thought seems to be taking symbolic cues from this rather irreverent tradition as well. The symbolisation of the creature emerges through the tension between a reverence towards, and a desire to revise, the Christian theological tradition. On a second, related point, I argue that the romantic train of reference pulls Whitehead's discussion of creatureliness into a discussion of *value*, leaving us, perhaps, with creatureliness itself as a symbolic function that insistently communicates the enjoyment of being actual. Thus it is only through the tension between reverence and revision – both a sense of respect and a desire to reinvent – that Whitehead is able to develop this political theology of enjoyment. In this sense, then, Whitehead is both secularising theology and theologising what might otherwise be something secular.

Romantic Creatures

In my own research into literary, theological and philosophical manifestations of the figure of the creature,[3] I have been consistently struck by the almost libertine nature of creatureliness in the British intellectual context. Given that the creature is, theologically, the mortal flip side of the immortal creator, many articulations of creatureliness in Western literary and philosophical contexts tend to focus (almost exclusively) on the finitude and vulnerabilities in creaturely life, presenting the creature as – in essence – a figure that is primordially cringing in a posture of submission or subjection. Thus in one sense the figure of the creature serves, in political theology, as the illumination of ultimate or abject powerlessness.

To be sure, there are British thinkers and writers whose creatures are, fundamentally, cringing in their vulnerability. In the work of Charles Dickens, for example, we often see the figure of the creature crop up to underscore a relation of unequal status. This is, for instance, the moniker that the beleaguered and abused Smike adopts for himself in *Nicholas Nickleby*. 'In the churchyard we are all alike', he tells Nicholas mournfully, 'but here there are none like me. I am a poor creature, but I know that'. Nicholas attempts to eviscerate some of Smike's pessimism, but still does not eliminate the diminutive status of the creatureliness that Smike has claimed for himself. 'You are a foolish, silly creature', Nicholas responds with some cheer. 'If that is what you mean, I grant you that'.[4] Indeed, this more particular figure of the 'poor creature' – innocent, yet pathetic – appears over and over again throughout the course of this tale. Again and again, characters declare themselves – in their subjection to the misery of their circumstance – to be poor creatures. Arguably, a political theology of submission or subjection continues to shape creatureliness in this narrative.

But I would argue that we begin to see in the work of those such as philosopher Mary Wollstonecraft[5] – particularly in her deployment of the term 'fellow creature' – the emergence of a creatureliness that is not invulnerable, but which itself seems to have the intrinsic capacity to push back against submission or subjection. Wollstonecraft's feminist treatise *A Vindication of the Rights of Woman* (1792) is loaded with creatures. She appropriates the Enlightenment discourse of reason to amplify a critical resonance between men and women – they are both, she charges, thinkers who seek the life of the intellect. They are both 'rational creatures'. Women and men, she argues throughout the text, are 'fellow creatures' – their faculty of reason being, in large part, what binds them in this fellowship. Her primary aim in the treatise, she declares, is to 'first consider women in the grand light of human creatures, who, in common with men, are placed on this earth to unfold their faculties'. It is only *after* she has made their fellowship in creatureliness quite clear, she clarifies, that she will point to the particular 'designations' that make women into women.[6] Thus, by calling men and women fellow *creatures*, she builds connective tissue between them while also preserving the gender distinction between disparate forms of creatureliness.

Wollstonecraft ultimately argues that becoming cognisant of this creaturely fellowship should lead a man to feel 'drawn by some cord of love to all his fellow creatures', including (perhaps above all) those female creatures who share his public and private spaces.[7] She appeals to the benevolent, love-riddled authority and function of this theological figure, seeking to make use of it as a political term that questions male supremacy and enjoins his real fellowship. Thus, the fact that all creatures are subjected to the benevolence of this all-powerful and immortal love is how Wollstonecraft seems to be preserving a kind of reverence for this symbol of the creature and its attendant political theology. But, on the other hand, she also seems to be revising it. She seems to see, in this politically theological figure, the power and capacity to push back against a social injustice. The figure of the creature, as a symbol, seems to have the powers and capacities to revolt and generate change.

It is impossible to say how much the work of Wollstonecraft's daughter, Mary Shelley, is caught up in the symbolic train of reference that Wollstonecraft is conducting here. What we do know is that Mary Shelley was familiar with her mother's work, and that the figure of the creature played a critical role in her novel *Frankenstein* – Victor Frankenstein's creation never receives a proper name, and always remains simply 'the creature'. The entire tale is a test of the fellowship between the creaturely creator Frankenstein and his creature. I will not make too much of the parallels between this narrative of a creaturely creator and Whitehead's God, who is also a creature transcended by the creativity that qualifies him. For this particular revision of the cosmological scene was common in the work of other British romantics – including Mary Shelley's husband, Percy. Percy Shelley was also familiar with Wollstonecraft's work, and may arguably have been caught up in this train of creaturely reference that I have begun to describe. Moreover, it is clear that Percy Shelley made some impact on Whitehead's work and thought, judging from the cautious approval Shelley receives as part of the 'romantic reaction' in *Science and the Modern World*.

Percy Shelley did not make much of Wollstonecraft's phrase 'fellow creatures', but the passages in his work where we do see it appear are illuminating. In his first large poetical work, *Queen Mab: A Philosophical Poem, With Notes* (1813), which was later revised into *The Daemon of the World* (1816), Shelley is quite forthright about his desire to revise the creation cosmology set forth by (especially in this case) Christian monotheism. This is a poem that expresses distaste for the despotism of the creator, and a faith in the abiding goodness of creatures in their own natural environment. The reaction that Shelley attempts to generate in this work is a moral revolution – one that asserts the presence of morality in nature, rather than having it doled out from a god on high. The legend of Queen Mab is of a tiny fairy creature who works parasitically in human life through people's dreams. In this narrative, Queen Mab helps to examine the dreams of humanity and infect them with new drives – in this case, the dream of a new kind of creaturely life.

Shelley criticises the 'despotic decrees'[8] of theology's 'terrible Deity'.[9] In one of his lengthy notes on this work, Shelley expresses his scepticism as to the existence of this deity by arguing that, if this god has the power that theology has long argued, 'no man would have had the effrontery to impose on his *fellow creatures*, in the name of the Deity, or to interpret his will according to his own fancy'.[10] Fellow creatures, here, function on one level as a moral critique of the pious – Shelley is using a term of the faith to critique what some actors have done in the name of the faith, against fellows in the faith. This is intended to expose, it would seem, a kind of hypocrisy. And yet, at the same time as Shelley is rejecting the political theology of the despotic creator, the figure of the creature endures – limned with a kind of power that generates a critique *of* that political theology. The creator dissipates in a way that the creature does not. There is a bit of rebellion in the creature.

This sort of reaction against, or revolution within, cosmology was not unique to the Shelleys, of course. Paul Cantor has argued that the British romantic tradition exemplifies – almost uniformly – what he calls a 'sympathy for the devil', raising 'doubts about the established order of the gods', questioning the gods' motives for creation, and celebrating the 'nobility' of figures who resist their power.[11] Cantor also argues that this unorthodox approach is most flagrant in William

Blake and Percy Shelley.[12] Celebrating the figure of the creature is, to be sure, not quite the same as celebrating the devil. But there is, I think, something of a rebellious (thus, to follow Shelley, a noble) aspect to the creature who symbolically emerges from the trains of reference I have been gesturing towards.

To what extent did the romantics' almost reverential, yet deeply revisionary approach to the creature–creator relation influence Whitehead's symbolisation of the creature? It may not be possible to determine with absolute certainty, though it is clear that there are some structural overlaps: the figure of the creature endures, while the figure of the creator (and the despotism associated with it) seems to fade into the background. Indeed, Whitehead – like Shelley – was concerned about the political theology of the despotic deity. In both Christian theology and in what he called in *Adventures of Ideas*, 'the patterns of Christian emotion' there have 'survived throughout history the older concept of a Divine Despot and a slavish Universe, each with the morals of its kind' (AI 26). Strikingly, it would appear that Whitehead is still thinking about the practical and political effects of the despotic deity against which Shelley once railed. Whitehead's commentary on the 'romantic reaction' in *Science and the Modern World* indicates that he found work like Shelley's useful for (among other things) its protest against the helplessness of creatures – in the face of 'irresistible grace' or the 'irresistible mechanism of nature' – that he saw in Protestant Calvinism, Catholic Jansenism and his contemporary science (SMW 75). He seems to find, in these romantics, the illumination of a kind of power – one that we might (although, in this work, Whitehead does not) call creaturely. It does not seem absurd to imagine that the figure of the creature achieves symbolic importance in Whitehead's work in large part because of the particular sort of symbolic importance it had in the context of the British romantic tradition.

These connections I am highlighting admittedly seem loose and potentially superficial; like symbolism, they might be described as unessential. But it seems unlikely to me that there is an exact science to symbolic analysis. If a symbol is a figure that has been caught up in trains of reference, what evidence for symbolic influence needs to be established beyond the illumination of a probable thought train, mediated through forms of direct intellectual contact? Perhaps, on the other hand, this does nothing but amplify the anxieties of influence. This possibility aside, I do think there is a kind of intellectual benefit to seeking out the influences exerted by these trains of reference: we might actually determine, with more precision and clarity, how a given symbol is functioning in context.

Drawing out these trains of reference surrounding the symbol of the creature, I have attempted to invoke a tradition of sorts – a practice of symbolic interpretation – that may well have influenced Whitehead's decision to strategically and symbolically deploy the figure of the creature in his own work. Whitehead seems to maintain his own reverence for the romantic reverential revisions of the figure of the creature. Whitehead's creature, like Percy Shelley's (as well as, arguably, the creatures of Mary Shelley, Wollstonecraft and others) is a creature who is at least partially free of divine tyranny and despotism, possessing its own powers. But Whitehead also revises this romantic tradition, in a sense, by bringing a different (and slightly more orthodox) god back into the cosmos. He is caught up in these romantic trains of creaturely reference, then, but only to a degree.

If we are, however, willing to see the continuity in this train of symbolic reference – to see the creature as reverentially revised out of *both* the Christian theological tradition *as well as* the British romantic literary tradition more particularly – then I think it becomes possible to catch a better glimpse of what specific function the creature-as-symbol might be playing in Whitehead's work. Indeed, this might even give us a clearer sense of what the generic creature – as figure, as symbol, as conceptual category – might *be* or might *do* when its relation to a creator figure becomes non-essential. This may help us to think, in other words, about what a creature becomes when it stands in fellowship with other creatures, but 'on its own' when it comes to creators. This is not a secularised creature, per se, given the fact that the creature's theological genealogy inevitably perdures on some basic level. But it is a creature with a distinct political theology – one that contests a political theology of despotism and promotes, perhaps, one oriented around enjoyment.

Creatures and Enjoyment

I want to return, then, to the initial line of questioning that animated this piece to begin with: if the creature is more than just a simple cognate for the actual entity, or a mere shadow of the actual entity, then what conceptual function might this category be serving in Whitehead's thought? If we entertain the line of thought I have been advancing – that Whitehead's creature-as-symbol is part of a romantic train of reference – then I think the function of the creature, in this train of reference, becomes value-oriented. The creature symbolises something value-related.

As Whitehead notes of the romantic reaction in *Science and the Modern World*, he is interested in this moment of thought not *only* because it was 'a protest on behalf of the organic view of nature' but *also* because it was 'a protest against the exclusion of value from the essence of matter of fact' (SMW 94). This reaction in thought and sensibility was a protest against the exclusion of value from actuality. What is value? It is, he notes here, 'an element which permeates through the poetic view of nature' (SMW 93). There is, then, something deeply aesthetic in value – an element that the poetry of the romantic reaction helps to preserve and defend. To open the symbolic train of reference leading out of the romantic reaction and into Whitehead's work, then, leaves our interrogation charged with *value*.

In *Religion in the Making*, Whitehead argues that value is not something that can be considered apart from, or independently from, actuality. 'Value is inherent in actuality itself.' Value, as something inherent in actuality and actual entities, is affective. It has 'an emotional tone' (RM 87). This emotional tone is an immediate experience, an ultimate physical fact, that results in enjoyment. As John Cobb Jr and David Ray Griffin have noted, this does not mean that actuality is pleasurable, as opposed to painful. Nor does it set up a contrast between entities that enjoy and entities that do not.[13] As Whitehead notes in *Modes of Thought*, the enjoyment of actuality is not aimed purely and simply towards goodness. 'Our enjoyment of actuality is a realization of worth, good or bad. It is a value experience. Its basic expression is – Have a care, here is something that matters!' (MT 116). Enjoyment generates and provokes attention in a very basic way. Value is essentially what calls entities to the attention of entities. As the facet of actuality that leaves these actual entities charged with enjoyment, value is also that aspect of reality that drives attention and attentiveness.

There is nothing religious about value, in the orthodox sense. Value is, Whitehead is very clear, not something that is god-given or externally mediated from on high. Value is entirely part of the 'self-interest' (RM 87) of the 'self-creating creature' (RM 89). But there is certainly something religious about value, if we think of religion in the Whiteheadian sense. Ultimately, for Whitehead, there would not be this sort of ultimate enjoyment without the function of God. This enjoyment of actuality requires a kind of 'perceptive fusion' that keeps it from becoming nothing but a 'confusion neutralizing achieved feeling' (RM 91). There is something, then, that allows value itself, the enjoyment of actuality, to become a higher-order level of feeling, something beyond an affect that serves simply to neutralise confusion. Confusion, neutralised, is a low-grade level of order. This is where religion, and God more specifically, function. God is the 'ordering entity' (RM 91) that presents ideal harmony to the world through the mental pole of actual entities. This order is, first and foremost, aesthetic. But it is also moral – to the extent that morality is 'merely [a] certain aspect of the aesthetic order' (RM 91). This 'harmony of apprehension', this perceptive fusion, is how God 'issues into the mental creature as moral judgment' (RM 105). The enjoyment of actuality is, then, a complex perceptive fusion linked to the mental pole of actual entities charging them, aesthetically and morally.

The realm of value – this enjoyment of actuality – contains a religious element or dynamic. What is religious about this, essentially, is the perceptive fusion in the mental pole of actual entities that is experienced as harmonious, ideal, ordered. The form of attention in value calls actual entities to attend not to divine commands, but to what matters – to matter, quite simply. It is, perhaps, an attention that harmonises between the mental and physical poles, that illuminates the order that plays out between actual entities and serves as a connective tissue in the shifting events of process. The value that is the enjoyment of actuality enlivens actuality to something – if not goodness in a simple sense, then to something harmonious playing out creatively. It enlivens actual entities to creativity, to the more complex harmonies within creativity.

As I opened this chapter, I made the suggestion that to call an actual entity a creature might simply be nothing more than to place the actual entity within the shadow of creativity – to remind us that the actual entity is conditioned by creativity. I want to return to this suggestion, but also to qualify it. After all of this, I do think that the creature – as symbol – seems to appear in Whitehead's work in order to complete this very basic task. Certainly Whitehead is not attempting to remind us that there is a creator who is infinitely bound to the creature. He has taken too many cues, I think, from the romantic reaction to turn back towards a political theology that might threaten with despotism. The creature does not depend on the creator figure in a manner that can be sketched out in such sharp relief. But there is, still, something religious (something evoking divinity) in the value experience that Whitehead calls the enjoyment of actuality – that harmony that keeps actuality attentive to what matters, creatively. The figure of the creature, as a symbol, names and thus symbolises that enjoyment of being actual. To speak of the creatureliness of actuality is a poetic gesture that symbolises the enjoyment of actuality – its potential harmoniousness. The actual entity is *also* a creature because there is a perceptive fusion within actuality, bringing it into a higher-order level of feeling, driving it to attend to what harmonises – what ropes it into aesthetic and moral ideals. The creatureliness of the actual entity seems to mark it with that somewhat crooked tie back to what Whitehead understands to be religion.

The presence of the creature-as-symbol in Whitehead's work is testimony to his penchant for both reverence and revision. The creature exposes, it seems to me, a certain kind of reverence for the Christian theological tradition as well as the romantic tradition in British literature. And yet Whitehead's strange and particular use of this symbol also exposes his willingness – or need – to revise these symbolic inheritances that he also gladly receives. The figure of the creature seems to qualify as a symbol in Whitehead's work primarily because of its seemingly superficial quality – it is difficult to discern what the function of this figure *is* in Whitehead's work, until we consider the possibility that it is being used as symbol to tie it back into the trains of reference that constitute this symbol's past and history. When we think of the creature as a symbol we can perhaps see with more clarity the sort of 'sociological statesmanship' Whitehead was involved in – the particular manner in which he attempted to mediate between worldviews, cosmologies and epochs of thought by toying with his symbolic inheritance. We also see, I think, through the open doors of a boxcar, a simple symbol as it travels along the tracks, still riding on an arbitrary train of reference that continues to head in new directions.

Notes

1 'The individual fact is a creature, and creativity is the ultimate behind all forms' (PR 20).
2 '. . . every actual entity, including God, is a creature transcended by the creativity which it qualifies' (PR 88).
3 My dissertation was submitted in August 2014, and was titled 'Dream of the creature: religion, ethics, and interspecies kinship'. One chapter of the dissertation involved an investigation of the etymological roots of the term 'creature', as well as an exploration of its uses in biblical and theological contexts. Another chapter explored the dynamics of power and submission in political theologies of creaturely life. This research

is presented in a forthcoming book tentatively titled *Creaturely Lives: Between Religion and the Secular.*
4. Dickens, *Nicholas Nickelby*, 275. Accessed via Google Books, 25 September 2014.
5. I also see traces of this form of potent creatureliness, with its own capacities and not exhaustively defined by subjection or submission, in the work of the seventeenth-century philosopher Anne Conway. Conway was also interested in the powers and capacities of non-human creatures, notably arguing that animals have their own forms of morality. There may, then, be something in this changing valuation of creatureliness that is linked to changing perspectives on animal life as knowledge became increasingly reoriented around animal biology. I have also been entertaining the possibility that there is some sort of gendered dynamic to the emergence of this more robust and lively form of creatureliness in intellectual thought. But this is not an argument I am entirely prepared to advance, at this point.
6. Wollstonecraft, *A Vindication of the Rights of Woman*, 33. Accessed via Google Books, 12 February 2012.
7. Wollstonecraft, *A Vindication of the Rights of Woman*, 205.
8. Shelley, *Queen Mab*, 94. Accessed via Google Books, 28 September 2014.
9. Shelley, *Queen Mab*, 95.
10. Shelley, *Queen Mab*, 95, emphasis added.
11. Cantor, *Creature and Creator*, ix.
12. Cantor, *Creature and Creator*, ix.
13. Cobb Jr and Griffin, *Process Theology*, 16–17.

Bibliography

Cantor, Paul A., *Creature and Creator: Myth-Making and English Romanticism* (Cambridge: Cambridge University Press, 1984).

Cobb, Jr, John B., and David Ray Griffin, *Process Theology: An Introductory Exposition* (Philadelphia: Westminster Press, 1976).

Dickens, Charles, *Nicholas Nickelby* (New York: Hurd and Houghton, 1867).

Shelley, Percy Bysshe, *Queen Mab, With Notes*, third edition (New York: 'Beacon' Office, G. Vale, 1842).

Whitehead, Alfred North, *Adventures of Ideas* (New York: The Free Press, [1933] 1967).

Whitehead, Alfred North, *Modes of Thought* (New York: The Free Press, [1938] 1968).

Whitehead, Alfred North, *Process and Reality: An Essay in Cosmology*, corrected edition, ed. David Ray Griffin and Donald W. Sherburne (New York: The Free Press, [1929] 1978).

Whitehead, Alfred North, *Religion in the Making* (Cambridge: Cambridge University Press, [1926] 2011).
Whitehead, Alfred North, *Science and the Modern World* (New York: The Free Press, [1925] 1967).
Whitehead, Alfred North, *Symbolism: Its Meaning and Effect* (New York: Fordham University Press, [1927] 1985).
Wollstonecraft, Mary, *A Vindication of the Rights of Woman* (London: T. Fisher Unwin, 1891).

6

Ren and Causal Efficacy: Confucians and Whitehead on the Social Role of Symbolism

HYO-DONG LEE

The Confucians in East Asia have always dreamed of holding human communities together and constructing well-functioning polities in and through the binding and harmonising power of rituals. Underlying their trust in the power of rituals is the notion that rituals constitute symbolic articulation and enhancement of our affective responses to the conditions of embodied relationality and historicity in which we always already find ourselves. This Confucian theory of rituals resonates with Whitehead's theory of symbolism, insofar as the latter advances a fundamentally relational ontology of the subject by highlighting the hitherto neglected epistemological notion of perception in the mode of causal efficacy. In this chapter I will attempt a comparative analysis of the two theories in order to gain a fresh cross-cultural perspective to appreciate Whitehead's implied critique of the modern liberal social theories that are based on a view of human beings as atomised individuals who rationally consent to enter society. My thesis is that both the Confucians and Whitehead offer theories of symbolic action predicated on radically relational understandings of the self, with Whitehead underscoring the historicity of causal connections among the high-grade organisms and the Confucians emphasising the primordiality of affective relations, especially within the context of the human family.

Whitehead on Symbolism

At the end of his *Symbolism: Its Meaning and Effect*, Alfred North Whitehead offers the following reflection on the social function of symbolism:

No elaborate community of elaborate organisms could exist unless its systems of symbolism were in general successful. Codes, rules of behaviour, canons of art, are attempts to impose systematic action which on the whole will promote favourable symbolic interconnections . . . The object to be obtained has two aspects; one is the subordination of the community to the individuals composing it, and the other is the subordination of the individuals to the community. Free men obey the rules which they themselves have made. Such rules will be found in general to impose on society behaviour in reference to a symbolism which is taken to refer to the ultimate purposes for which the society exists. (S 87–8)

The social role of symbolism, in other words, is to provide space for the freedom of individuals while ensuring social cohesion by guiding and channelling the exercise of freedom to serve the ultimate purposes of the society of which the individuals are members. Our 'vast system of inherited symbolism' (S 73), Whitehead avers, reconciles the requirements of social preservation and 'the contrary stimulus of the heterogeneity derived from freedom' (S 65) first by 'adding emotion to instinct' and secondly by providing 'a foothold for reason' through 'its delineation of the particular instinct which it expresses' (S 70). To put in another way, symbolism enhances and amplifies our instincts with emotions and at the same time enables a rational discernment and articulation of them. One can say with Whitehead that symbolism affectively intensifies our instinctive bonds with one another, preserving social unity and cohesion, while allowing the reflective subjectivity and agency of the individual to play a pivotal role in that process of intensification, guaranteeing that the instinctive bonds thus strengthened are not blind, but beholden to 'fluctuating apprehension of the basis of common purposes' (S 73).

Here it is worthwhile to note that the instinctive social bonds of which Whitehead speaks are much more prominent in the inorganic world and the societies of more primitive living organisms, and that as a result those societies enjoy much longer spans of existence than the more developed types of living communities: 'So far as survival value is concerned, a piece of rock, with its past history of some eight hundred millions of years, far outstrips the short span attained by any nation' (S 64–5). In fact, the most successful examples of community life are found in the inorganic world, 'among societies of active molecules forming rocks, planets, solar systems, star clusters' (S 82), which are 'kept together by the blind force of instinctive actions' (S 68). The emergence of life is, then, to be conceived as 'a bid for freedom on the part of organisms, a bid for a certain independence of individuality with self-interests and activities not to be construed purely in terms of environmental obligations' (S 65). Despite the deleterious effect it has had on social preservation through its introduction of heterogeneity, this bid for freedom – especially among the more developed forms of life – has successfully connected the 'self-interests and activities' of the individual with the 'common purposes' of a society via an alternative and more comprehensively effective mechanism for social cohesion, namely symbolism (S 65–6).

One of the distinct features that can be observed in Whitehead's discussion of the social function of symbolism is the pivotal role which his theory of perception plays in it. According to his formal definition of symbolism, 'the human mind is functioning symbolically when some components of its experience elicit consciousness, beliefs, emotions, and usages, respecting other components of its experience' (S 7–8). Whitehead calls the former set of components of experience 'symbols', the latter set the 'meaning' of the symbols, and the transition from the symbol to the meaning 'symbolic reference' (S 8). More specifically, symbolic reference is a 'synthetic activity' that works by connecting the two modes of perception – 'presentational immediacy' and 'causal efficacy' – in such a manner that one functions as the symbol and the other provides the meaning of it (S 17–18). While there are no components of experience that are only symbols or only meanings, the usual direction of symbolic reference runs from the less 'primitive'[1] components as symbols to the more primitive components as meanings (S 10). Here the critical claim made by Whitehead is that perception in the mode of causal efficacy is the more primitive component of experience, and that therefore symbolic reference runs from presentational immediacy to causal efficacy. Causal efficacy is the experience of the presences or powers which we perceive incontrovertibly as affecting our bodies directly but indistinctly, and to which we respond so immediately that their efficacy always already belongs to the past conditions that have given rise to our present state (S 48). Further, this experience of causal efficacy is shot through with emotional tones, as our responses to these vague presences are closely entwined with feelings of 'anger, hatred, fear, terror, attraction, love, hunger, eagerness, massive enjoyment', depending on the specific character of the externality impressing upon us (S 44).[2] This affectively rich experience of causal efficacy is so certain that the practical responsiveness of the organisms involves little hesitance or irresolution, leading the organisms to anticipate firmly the conformation of the immediate future to their present activity (S 42).

In the perceptive mode of presentational immediacy, by contrast, the external environment appears as a community of mutually distinct actual things, but only by the mediation of qualities such as colours, sounds, tastes and so on, usually named sense-data (S 22). This means that, although presentational immediacy offers us vivid perceptions of the world of actual things precisely delineated and spatially related to one another, such perceptions can be 'barren', for we cannot directly connect the perceived qualities with any intrinsic characters of the actual things (S 24). In fact, presentational immediacy in its pure form is a highly abstract operation found in a meaningful way only in the experience of a few high-grade organisms (such as humans) (S 23). Symbolic reference counters this once-removed character of presentational immediacy from the immediate practical exigencies of the world of actual things by referring the perceived qualities (including the perceived emotional impact on one's body) as symbols to the perceived efficacious presences as the meaning of the symbols.[3] Symbolic reference provides specific delineation and enhancement of the vague perceptions of causal efficacy, while anchoring the distinct perceptions of presentational immediacy in the world of actual things.

Whitehead's theory of symbolic reference is meant to counter the modern European epistemological tradition, represented by Hume, that treats presentational immediacy as the only type of perceptive experience and excludes from it any demonstrative factors linking it to a world of actually existing, efficacious things (S 34).[4] Whitehead rejects this view as assuming a wrong understanding of experience predicated on a fundamentally mistaken atomic conception of the self (S 35). His account of symbolic reference reveals the 'relational ground' (S 35) in which the self comes into being first and foremost through its reception and active incorporation of the relevant, causally efficacious elements of other actual things of the past as components of its own experience. And what underlies this radically and primordially relational account of our experience is an earlier version of the relational process ontology of the self fully articulated in *Process and Reality*:[5] all actual occasions are self-determining and self-creative subjects-agents, partially obligated to and determined by the past of one another, yet also free to propose and realise new shapes of their becoming. Among the more complex societies of actual occasions, particularly the most developed communities of life with both sense-perception and conceptual analysis, the symbolic capacity to enhance, articulate, and re-imagine the fact of causal obligations in reference to some visions of shared becoming testifies to their genuine self-creative freedom.

This relational ontology of the self presupposed in Whitehead's theory of symbolism explains his criticism of Locke's social contract theory as 'a baseless historical fiction' (S 72). Like any actual occasions of experience or societies of them, human beings are not atomic selves primordially independent of one another and only derivatively related. They do not enter relations through their exercise of the high-grade mental functions of knowledge and will, such as a rational and free decision to sign a social contract. Human beings always already find themselves in the conditions of embodied relationality and historicity, causally determined by the weight of the past others, yet free to re-imagine and reshape that heritage by means of alluring symbolisms of common living. On this point, the Confucians would certainly agree.

Confucianism and Ritual[6]

In the year 1791, Paul Yun Ji-chung and his cousin James Gwon Sang-yeon, both Korean Confucians who had converted to Catholic Christianity, burned the ancestral tablets of their family and refused to continue practising the traditional Confucian ritual of ancestor veneration. Further, upon the death of Yun's mother they rejected the traditional Confucian rites and held a Catholic funeral mass instead. The communal response to the stir they had created was swift. The two were arrested by the staunchly Neo-Confucian government of the Joseon Dynasty, and after refusing to renounce their Catholic ways they were executed by beheading.[7]

The severity of the method of punishment – which was normally reserved for the crime of treason when the convicted belonged to the ruling class of Confucian literati – says something about the level of shock and alarm the two martyrs had caused in governing circles. This is particularly evident when compared with an incident six years earlier in which a group of Confucian scholars were arrested in a police raid while gathered to study Catholic doctrines – they were soon released with a mere rebuke.[8] Whereas merely having ideas at variance with the established Neo-Confucian orthodoxy or even teaching them was more often than not met by an attitude of annoyed toleration, any deviation from the time-honoured Confucian rituals was regarded as a grave threat to the very fabric of human society. A glance at the history of the Joseon Dynasty will show that political factions almost never won or lost a battle for power strictly on the basis of the reputed orthodoxy of their doctrines. If, however, your party's interpretation of how to carry out the ritual of mourning for the deceased queen mother was rejected by a king who favoured the other party's interpretation, you and your comrades would be thanking the king for dispatching you into the harsh life of an exile and not sending you a gift of poison to drink. You may read the strange classics of so-called Western learning (Catholic Christianity) to your heart's content and learn its uncouth and barbarian catechisms by heart, but if you mess with the rites of the sages you will have to face the consequences.

Why this obsession with rituals in Confucianism? Rituals, the Confucians believe, are performative articulations of the humanity of your heart-mind (心, *xin*). In other words, without rituals, your heart-mind is not human. For example, the rites of honouring and venerating your ancestors, including your parents, function as testimonies to and touchstones for your filial piety (孝, *xiao*) – the single most important articulation in practice of the cardinal Confucian virtue of humanity (仁, *ren*), that is, the very thing which makes a human being human. Filial piety is humanity as it is present between parents and children in the emotional tone of affection (親, *qin*), and the Confucians regard the parent–child relationship as the most fundamental of all human relations.

According to Confucius or Kongzi (孔子), the revered founder of the Confucian Way (道, *dao*), humanity may be defined as the integrity of a guileless self (忠, *zhong*) with a capacity for empathetic response to – or sympathetic understanding of – others (恕, *shu*).[9] In other words, Kongzi puts at the heart of his definition of humanity an ideal of selfhood that is open, empathetic, relational and all-embracing.[10] He views such an ideal selfhood as both being formed by and finding its bodily enactment in rituals, the view captured by his famous dictum: 'Overcome the [self-centred individual] self and return to ritual'.[11] By 'rituals', he does not mean any rituals, but the ones permeated by a spirit of mutuality and reciprocity. Such rituals are productive of social concord and harmony, for they both manifest and help bring into being empathetic selves capable of being completely in tune with one another, that is to say, understanding one another sympathetically and hence always appropriate – right (義, *yi*) – in their ritual responses to one another. His lifelong attempt to renew the accumulated ritual tradition of the declining culture of the Zhou Dynasty was grounded in his conviction that the spirit of mutuality and reciprocity generated by empathetically related selves constituted the forgotten essence of the Zhou rituals of divine–human and inter-human interactions. He tried to unlock and retrieve that essence through a programme of classical, literate learning and ritual self-cultivation.

Kongzi's idea of a radically open, empathetic and relational selfhood and its ritual articulation finds a further anthropological and psychological development in Mencius or Mengzi (孟子), historically the second most important figure within the Confucian tradition. According to Mengzi, humanity or *ren* names the human nature with which everyone is born – 'nature' (性, *xing*) here being the spontaneous course in which a life-form completes its development when nurtured and not obstructed.[12] As such, human nature can be said to have been decreed or endowed by heaven as a cipher for nature as *physis*, that is, ceaseless generativity (生生, *shengsheng*).[13] Understood as *ren*, human nature can be said to point to the 'seed' of a radically open, empathetic and relational selfhood that is in all of us humans as the core human potential to be developed fully. The 'seed' of humanity consists of four 'sprouts' present within every human being, namely, the four good feelings of sympathy and benevolence, shame and dislike, deference and compliance, and approval and disapproval. As diverse relational articulations of human nature understood as empathy, they culminate in the Four Virtues of benevolence, rightness, ritual propriety and wisdom (人義禮智, *renyilizhi*).[14] Mengzi describes the growth of these so-called four sprouts (四端, *siduan*) in terms of the bodily cultivation of one's *qi* (氣), namely, the vital psychophysical energy of the universe of which everything is made up.[15] This bodily self-cultivation relies on both the spontaneous issuing forth of the four sprouts of empathetic feelings and the deliberative capacity (思, *si*) of one's heart-mind as the faculty of judging the relational and situational appropriateness of our various feelings, inclinations and actions.[16] Driven by the four sprouts and guided by the deliberative capacity, the bodily self-cultivation accumulates right – that is, empathetic and measured – moral and ritual responses to others in diverse relational contexts.[17] The accumulation of right moral and ritual responses enables ever deeper and more resonant connection between one's bodily psychophysical energy and that of others. It progressively expands the boundaries of one's self until it comes to encompass the entire universe in empathy by fully resonating with the cosmic psychophysical energy filling heaven and earth.[18]

As the dominant school in the history of the Confucian tradition from the eleventh century to the nineteenth century CE in East Asia,

Neo-Confucianism took up the Mencian development of Kongzi's idea of humanity and its ritual articulation cum formation and formulated a moral-psychological account of the human being that had at its centre what may be called a relational ontology of affectivity. The representative Neo-Confucian moral psychology advanced by the so-called Cheng-Zhu school starts with the dictum 'The heart-mind unites nature and feelings' (心統性情, xintong xingqing).[19] The heart-mind here functions as a synecdoche for the self, understood to consist in a union of pattern (理, li) and psychophysical energy (氣 qi), pattern here being the metaphysical structure of reality that is logically and ontologically prior to psychophysical energy, yet is always found intertwined with and dependent on the latter for its creative dynamism. According to this Neo-Confucian metaphysical development of the Mencian moral psychology,[20] nature (性, xing) stands for pattern as it is 'incarnate' in each thing as individual coalescences of psychophysical energy, while feelings (情, qing) designate nature's most primordial manifestation as the relationally patterning/structuring force in concrete relational contexts, which is powered by the spontaneous dynamism of psychophysical energy.

In the case of the four sprouts of empathetic feelings, they are initial, embodied, affective responses of the heart-mind to others which follow without deviation the dictates of the human nature within. As such, they retain the radically open, empathetic and responsive state of 'equilibrium' (中, zhong) characteristic of the heart-mind completely in sync with the 'mandate' of human nature prior to its 'awakening' and issuing forth in affective responses. Hence, as empathetic, other-oriented, and therefore measured feelings, they structure harmonious (和, he) relations as they issue forth in various relational contexts in the form of ritually correct actions.[21] The embodied, affective responses of the self may, however, get swept up in the spontaneous and unruly dynamism of psychophysical energy and deviate from the dictates of the human nature within. In such cases, they may lose the original equilibrium of the heart-mind and become unempathetic and unresponsive, that is, become excessive or deficient, being closed in upon themselves and not open to the call of relational exigencies. Discords, tensions and conflicts are products of such self-centred and relationally 'incorrect' feelings that express themselves in ritually inappropriate behaviours.[22]

The mandate of human nature calls upon the moral agency of the self – the so-called human heart-mind (人心, renxin) – to nurture relationally harmonious feelings, while bringing under control the non-harmonious ones by exercising intentional deliberation and judgement. When the human heart-mind succeeds in this task, it becomes identical to what is called 'the heart-mind of the Way' (道心, *daoxin*).[23] A repeated exercise of the human heart-mind's moral agency as the heart-mind of the Way over the long haul accumulates relationally correct moral judgements and ritual responses to such an extent that one's feelings are habitually conditioned to expresses themselves in ritually proper measures spontaneously, while one's moral judgement is perfected always to favour such spontaneous ritual responses. During the course of such successful bodily self-cultivation, the human heart-mind progressively transforms one's individual coalescence of psychophysical energy into a clearer, more open, balanced and responsive condition, and in so doing expands the boundaries of one's psychophysical energy beyond the self–other distinction, to encompass heaven and earth. This enables one to become a 'superior person' (君子, *junzi*) ready to join the ranks of fulfilled human beings, namely, the sages (聖人, *shengren*), who have an enduring and unwavering possession of the heart-mind of the Way. Learning the classics is crucial in this process of self-cultivation, precisely because it was none other than the sages of old who created the ritual tradition contained in the classics out of their unobstructed capacity to manifest and articulate the heart of humanity within them. The ritual learning of the tradition of the sages would be an unerring guide to properly measured – relationally correct – moral judgements and ritual responses in all situations imaginable and possible, as one's self progressively expands through a series of ever-enlarging concentric circles of relation, starting from the family and – through the local community and the state – ending at 'all under heaven' (天下, *tianxia*).[24]

The perennial socio-political dream of the Confucians has been rule by such superior persons or sages. As ritual masters whose meaningful symbolic actions manifest fully the animating heart of society, namely *ren*, superior persons and sages would radiate moral charisma that spontaneously draws people – and indeed all creatures under heaven – to them and to one another in peace and harmony. They would be like the North Star, which, though itself unmoving, pulls all other stars non-coercively into orbit around it to create celestial harmony, with all its beauty and grandeur.[25] They would be like Shun, the legendary sage-king, whose ritual action of facing the south, the auspicious direction, was enough to bring peace to his realm.[26] The Confucian programme of classical learning and ritual cultivation, which is in principle open to all regardless of their social station,[27] has aimed at educating rulers who are 'sage inside, king outside' (內聖外王, *neishengwaiwang*)[28] and the fellow ritual masters who can ably assist the sage-kings in the task of holding human society and the entire cosmos together and helping the myriad creatures in it flourish. Nowhere was this programme implemented with so much zeal, dedication and generation-bridging consistency as in Korea during the 500-year rule of the Joseon Dynasty, although it is another question whether the effort was ultimately successful. It is no surprise, then, that all hell broke loose when the early Catholic converts abolished the established symbolic action of honouring parents and ancestors – the very first, foundational component in the programme of ritual learning and ritual governance.

A Concluding Reflection: Affective Relations and Causal Connections

The two preceding expositions of Whitehead's theory of symbolism and the Confucian theory of ritual have shown, I hope, their agreement that society is held together largely by the glue of symbolic actions, and that symbolic actions are grounded in the primordial fact of our embodied relationality. Both of them put at the centre of their respective accounts of human experience a notion of the open self whose very coming into being is dependent first and foremost on its intercourse with other selves on some very instinctual and emotional levels. They also argue that the instinctual and affective exchanges between open selves are accompanied by the higher-level mediation of symbolic actions or rituals that amplify the efficacious power of such exchanges while turning them into objects of rational deliberation, moral assessment and willing consent. Their overlapping accounts of the symbolic mediation of human relations imply a rejection of any view of symbols as either empty linguistic embellishments or unifying sign-systems which, purely through the power of habit, prejudice or convention, are arbitrarily imposed upon people who are in reality – in the 'state of nature' – mutually unrelated autonomous individuals who should accordingly be brought together only by some kind of rationally entered social pact.

For Whitehead and the Confucians, the 'state of nature' is none other than society; and the most convincing demonstration of the factuality of this thesis lies in the socially most primary unit of parents and children, to give their shared claim a Confucian inflection.

This primary relational matrix for the becoming of the self is shot through with affectivity that is specified by the Confucians with the notion of parental and filial affection (親, *qin*). For the Confucians, the educational formation of a human being – *humanitas* – begins here, with the performative learning of symbolic actions to enhance, articulate and foster the affective quality of familial relations. The reason for such an educational approach lies in the belief that *qin* is the primordial and paradigmatic expression of *ren*, and as such provides a compelling reason to regard all human relations as extensions of familial relations.[29] A person who fails at the task of ritual self-cultivation within the familial context, unable to relate to one's parents or children with symbolically focused and measured *qin*, will not do better within the context of larger communities or the state, for the centre has been hollowed out from the ever-enlarging concentric circles of ritual learning. Hence, any claim made to the sagely ability to pacify the state or the world, without having first brought a ritual and affective harmony to one's family, will be greeted by the Confucians with a healthy dose of scepticism.

We may want to take our cue from the Confucians and pay more attention to ritual learning within the context of the family, to cultivate a rich tradition of family rituals where such has never received serious attention, and to recover it where it has been lost. After all, the family is emblematic, on the level of human organism, of the all-pervasive and in many cases overpowering causal weight of the past that Whitehead articulates. As such, for good or ill, this socially most primary unit arguably makes the greatest impact on the becoming of the human self from very early on. A lot is at stake in whether or not we are able to bring some order and harmony to the affective exchanges taking place in familial relations, and this task could be accomplished by cultivating this primary relational matrix in a deep reservoir of vibrant symbols beckoning us to the humanity – our fundamentally and empathically relational nature – within us. Many dominant responses to social problems in the Western world today treat them as tasks of social engineering and focus primarily on large-scale institutional measures – such as economic or public educational policies – which, although aimed at providing societal underpinnings for the well-being of family units, for the most part do not pay enough attention to their status as the primary site for nurturing the affective basis of human flourishing. In zooming in on the family, however, the Confucian point is not to regard the relations constituting the larger social units as extensions of the familial relations in the literal sense of the term, but to extrapolate imaginatively and analogically, through a process of symbolic transference, from the affective heart animating the latter in order to humanise the former.

Nevertheless, despite its spirit and intent, the Confucian emphasis on the family has historically disclosed the risk in this educational strategy – namely, the danger that ritual learning might be tied down by the power of the primary affective ties and not progress beyond the boundaries of the relatively homogeneous units of the family and the tribe to encompass ultimately all the differences in the cosmos. The conservatism and traditionalism that have historically characterised Confucian ritualism have exhibited a tendency to turn into nepotism and cronyism on numerous historical occasions. The ritual cultivation of analogical imagination to regard as wide a circle of human beings as possible as intimate members of one's own family has repeatedly stalled in the face of the perceived threat of heterogeneity lurking within the seemingly barbaric and inhuman heart-minds of others. We may nonetheless be able to check this exclusionary tendency with the help of Whitehead's consistent attention to the openings of novelty created by the active and inspired appropriation of the past on the part of subjects-in-becoming. If temporality is indeed historicity, as Whitehead contends; if time unfolds with the creative espousal of the always already patterned past to advance into the partially open future, rather than constituting mere duration or pure succession; then the great reverence in which tradition is held by Confucian ritualism need not mean a captivity to what is most 'familiar' because it is familial. As Whitehead argues, although a symbol itself may not change, it will have different meanings for different people through different times. Symbolism may need 'a continuous process of pruning, and of adaptation to a future ever requiring new forms of expression', and even an occasional revolution (S 61). That may have been what was intended by the great innovator, Kongzi, when he famously declared: 'Transmitting insight, but never creating insight, standing by my words and devoted to the ancients.'[30] To find oneself variously obligated to others in the unavoidable web of both familiar and unfamiliar causal connections, and to learn to name, assess and order those obligations by means of ritual interactions, always starting from what is most 'near at hand',[31] but constantly moving beyond it – such may be a Confucian–Whiteheadian wisdom on the kind of symbolism we need today.

Notes

1 Here the term 'primitive' is used in reference to the level of complexity found in organisms.
2 According to Rolf Lachmann, Whitehead's notion of causal efficacy is a precursor to his notion of physical prehension. Lachmann summarises Whitehead's account of causal efficacy found in *Modes of Thought* as follows: 'At the root of our consciousness lies the experience of the effectiveness of the various powers of the past – movement, activity, approach, distanciation, transition etc. – upon our current situation, the experience shot through with emotional tones such as anger, hatred, fear, love etc. and closely connected to our bodily experience. Depending on our bodily constitution we evaluate the effective powers as suitable/beneficial or disruptive/harmful, and these "meanings" get signaled through emotions' (211). Hence, for Whitehead, 'an emotion is an organically grounded meaning' (211). Symbolisation may accordingly be defined as 'a further continuation, articulation and elaboration of the bodily grounded and emotional meaning-production' (212). Lachmann, 'Alfred North Whiteheads naturphilosophische Konzeption der Symbolisierung'.
3 In presentational immediacy, the 'objects' of experience include bodily organs. In such cases the sense-data are called bodily feelings (S 22–3).
4 Whitehead's epistemological critique is presented fully in Part II, Chapter 8 of *Process and Reality*.
5 Thomas Hosinski correctly sees the perceptive mode of causal efficacy as corresponding to the initial physical prehension, but matches, confusingly in my view, the perceptive mode of presentational immediacy with 'the conscious grasping of a propositional feeling' without clearly explaining the role of conceptual prehension in the propositional feeling. The confusion is compounded by his relating of symbolic reference to the 'intellectual feeling' that integrates the data given in the propositional feeling with the data given in the original physical prehension. Hosinski, *Stubborn Fact and Creative Advance*, 118.
6 This section incorporates a heavily rewritten version of the account of Confucian moral psychology found in Chapter 2 of my book *Spirit, Qi, and the Multitude*.
7 For a concise history of the troubled early relationship between the Joseon Dynasty and the nascent Catholic Church in Korea, see Kim Seongtae, *Han-guk cheonju gyohoesa I* (History of the Korean Catholic Church I), 250–345. See also Cho Kwang, 'The Chosŏn government's measures against Catholicism', 103–14.
8 Choe Jaegeon, *Joseon hugi seohagui suyong-gwa baljeon*, 58–62.
9 See *Analects* (Lunyü), 4:15, where *ren* (仁) is construed in terms of integrity (忠, *zhong*) and sympathetic understanding (恕, *shu*). Zhu Xi, *Si shu zhang ju ji zhu* (Collected commentaries on the Four Books), 72.

10 Tu Wei-ming defines the Neo-Confucian notion of transcendence as such. See Tu, *Confucian Thought*, 51–65.
11 *Analects*, 12:1, in Zhu Xi, *Si shu zhang ju ji zhu*, 131.
12 Mencius, 6A6, in Zhu Xi, *Si shu zhang ju ji zhu*, 328–9. See also 6A7, 329–30; 6A8, 330–1; 2A6, 237–8; 4A27, 287.
13 Mencius, 7A1, 394. One should note that 性 (*xing*) and 生 (*sheng*) are cognate.
14 Mencius, 6A6, 328; 2A6, 238.
15 Mencius, 2A2, 230–2.
16 Mencius, 6A15, 335. The heart-mind is the faculty of reflecting on and judging the relative importance of our various appetites and moral inclinations. Without it, 'the senses simply yield to the attraction of what excites them and withdraw attention from everything else. We notice again that general assumption of his tradition, that moral thinking starts when the senses are already responding to stimulation from outside, and that its function is to choose between reactions in the light of the fullest knowledge' (Graham, *Disputers of the Tao*, 131–2). See also Mengzi, *Mengzi: With Selections from Traditional Commentaries*, xxxiv.
17 Such as the so-called Five Relations: parent–child, ruler–subject, elder–younger, husband–wife and friend–friend relations. Mencius, 3A4, 259.
18 See note 15. The cosmic psychophysical energy would be what Mencius calls 'vast, flood-like psychophysical energy' (浩然之氣, *haoran zhiqi*). I am partially borrowing Graham's translation of the phrase as 'the flood-like *ch'i*'. See *Disputers of the Tao*, 127.
19 The dictum was first coined by one of the earliest Neo-Confucians, Zhang Zai. I am using Wing-tsit Chan's translation found in his *A Source Book in Chinese Philosophy*, 517.
20 See the succinct description of the Neo-Confucian moral psychology by Michael Kalton in the introduction to *The Four–Seven Debate*, xxii–xxv. For the relationship among the heart-mind, human nature and feelings, see Zhu Xi, *Zhuzi yulei* (Conversations of Master Zhu, arranged topically), 1:89, 92, 94–5. For the role of intentional deliberation (意, *yi*), see 1:96. For Zhu Xi, desires are intensifications of feelings, and people have evil desires when their feelings become excessive and unbalanced to the point of being uncontrollable (1:93–4).
21 Mengzi's moral-psychological and philosophical-anthropological development of Kongzi's idea of humanity and its ritual articulation cum formation finds its parallel in the classical text of a Mencian heritage, the *Zhongyong* or *The Doctrine of the Mean*: 'What is mandated by Heaven is called nature; following the nature is called the way [道, *dao*]; cultivating the way is called education' (1.1); 'Before pleasure, anger, sorrow, and joy have arisen, it is called equilibrium [中, *zhong*]; after they have arisen and attained due proportion, it is called harmony [和, *he*]. Equilibrium

is the great foundation of the universe; harmony is the Way that unfolds throughout it' (1.4). Zhu Xi, *Si shu zhang ju ji zhu*, 17–18.

22 The precise relationship between the four sprouts of empathetic feelings (the Four Beginnings) and the rest (the Seven Feelings) was the pivotal question around which the celebrated Four–Seven Debates in Korean Neo-Confucianism unfolded. In the current discussion I am following arguably the most articulate exposition of the Neo-Confucian moral psychology cum philosophical anthropology by Yi I (Yulgok), a sixteenth-century Korean thinker-statesman. Yi I, 'Dap Seong Ho-won' (Reply to Seong Ho-won)', in Yi I, *Gugyeok Yulgok Jeonseo* (Complete translated works of Yulgok), III, 9. 35b–36a, 17–18 (for citations from *Jeonseo*, I give the volume number in roman numerals, the book number and the page number in the traditional format, and then the page number in the modern pagination). A good translation of these texts is given in Kalton's *Four–Seven Debate*, 115–16, 133–4.

23 Yi I, *Gugyeok Yulgok Jeonseo*, III, 9. 35b–36a, 17–18.

24 The succinct formulation of this self-cultivation process is found in 'Great Learning', 1.5. Zhu Xi, *Si shu zhang ju ji zhu*, 4.

25 *Analects*, 2.1 (Zhu Xi, *Si shu zhang ju ji zhu*, 53).

26 *Analects*, 15.4 (Zhu Xi, *Si shu zhang ju ji zhu*, 162).

27 *Analects*, 7.7; 15.38 (Zhu Xi, *Si shu zhang ju ji zhu*, 94, 168).

28 The term originally appears in the thirty-third (*tianxia*) chapter of the Daoist text *Zhaungzi*. Zhuangzi, *Zhuangzi ji shi* (Collected commentaries on Zhuangzi), 1064. It was taken over by Neo-Confucians to point to the Confucian ideal of rulership.

29 The familial paradigm in which all human relations and ultimately the cosmos as a whole are cast finds its supreme expression in Zhang Zai's *Western Inscription*: 'Heaven is my father and Earth is my mother, and even such a small creature as I finds an intimate place in their midst. Therefore that which fills the universe I regard as my body and that which directs the universe I consider as my nature. All people are my brothers and sisters, and all things are my companions. The great ruler (the emperor) is the eldest son of my parents (Heaven and Earth), and the great ministers are his stewards ... Even those who are tired, infirm, crippled, or sick; those who have no brothers or children, wives or husbands, are all my brothers who are in distress and have no one to turn to.' Chan, *A Source Book in Chinese Philosophy*, 497.

30 *Analects*, 7.1. I am using here the translation by David Hinton. Kongzi, *Analects*, 65.

31 I am referring to the celebrated Neo-Confucian concept of *jinsi* (近思) that encapsulates the practical and empirical orientation of Neo-Confucianism.

Bibliography

Chan, Wing-tsit, *A Source Book in Chinese Philosophy* (Princeton: Princeton University Press, 1963).

Cho Kwang, 'The Chosŏn government's measures against Catholicism', in Chai-Shin Yu (ed.), *The Founding of Catholic Tradition in Korea* (Fremont: Asian Humanities Press, 2002), pp. 103–14.

Choe Jaegeon, *Joseon hugi seohagui suyong-gwa baljeon* (The reception and development of Western learning in late Joseon) (Seoul: Handeul chulpansa, 2005).

Graham, Angus, *Disputers of the Tao: Philosophical Argument in Ancient China* (La Salle: Open Court, 1989).

Hosinski, Thomas E., *Stubborn Fact and Creative Advance: An Introduction to the Metaphysics of Alfred North Whitehead* (Lanham: Rowman and Littlefield, 1993).

Kalton, Michael, 'Introduction', in *The Four–Seven Debate: An Annotated Translation of the Most Famous Controversy in Korean Neo-Confucian Thought* (Albany: State University of New York Press, 1994).

Kim Seongtae, *Han-guk cheonju gyohoesa I* (History of Korean Catholic Church I) (Seoul: Han-guk gyohoesa yeon-guso, 2009).

Kongzi, *Analects*, trans. David Hinton (Washington, DC: Counterpoint, 1998).

Lachmann, Rolf, 'Alfred North Whiteheads naturphilosophische Konzeption der Symbolisierung', *Zeitschrift für philosophische Forschung*, 54:2 (April–June 2000), 196–217.

Lee, Hyo-Dong, *Spirit, Qi, and the Multitude: A Comparative Theology for the Democracy of Creation* (New York: Fordham University Press, 2014).

Mengzi, *Mengzi: With Selections from Traditional Commentaries*, trans. Bryan W. Van Norden (Indianapolis: Hackett, 2008).

Tu, Wei-ming, *Confucian Thought: Selfhood as Creative Transformation* (Albany: State University of New York Press, 1985).

Whitehead, Alfred North, *Process and Reality: An Essay in Cosmology*, corrected edition, ed. David Ray Griffin and Donald W. Sherburne (New York: The Free Press, [1929] 1978).

Whitehead, Alfred North, *Symbolism: Its Meaning and Effect* (New York: Fordham University Press, [1927] 1985).

Yi I, *Gugyeok Yulgok Jeonseo* (Complete translated works of Yulgok), ed. Hanguk jeongsin munhwa yeon-guwon jaryo josasil, 7 vols (Gyeonggi-do Seongnam-si: Hanguk jeongsin munhwa yeon-guwon, 1984–8).

Zhu Xi, *Si shu zhang ju ji zhu* (Collected commentaries on the Four Books) (Beijing: Zhonghua shu ju, 1983).

Zhu Xi, *Zhuzi yulei* (Conversations of Master Zhu, arranged topically), ed. Li Jingde and Wang Xingxian (Beijing: Zhonghua shuju, 1986).

Zhuangzi, *Zhuangzi ji shi* (Collected commentaries on Zhuangzi), compiled by Guo Qingfan, ed. Wang Xiaoyu (Beijing: Zhonghua shu ju, 1961).

7

Avoiding a Fatal Error: Extending Whitehead's Symbolism Beyond Language

SHERI D. KLING

In his foreword to Juan Eduardo Cirlot's text *A Dictionary of Symbols*, Herbert Read writes that every human 'is a symbolizing animal; it is evident that at no stage in the development of civilization has man been able to dispense with symbols'.[1] Though Alfred North Whitehead sees symbolism as being 'essential for the higher grades of life', he conversely states that humans are both attracted to and repulsed by symbols because 'hard-headed men want facts and not symbols' (PR 183; S 60). Because Whitehead saw symbols as uncontrollable, like 'wild vegetation' that could 'overwhelm' humanity, he believed that the advance of civilisation required that humans reject the superstitious symbols of their 'savage past' (S 60). Yet Whitehead acquiesces to the truth that despite all efforts to 'expel' it, symbolism – like nature – will 'ever return' and is 'inherent in the very texture of human life'. He continues, 'Mankind, it seems, has to find a symbol in order to express itself. Indeed "expression" is "symbolism"' (S 61–2).

Whitehead writes:

> the human mind is functioning symbolically when some components of its experience elicit consciousness, beliefs, emotions, and usages, respecting other components of its experience. The former set of components are the 'symbols', and the latter set constitute the 'meaning' of the symbols. (S 7–8)

Swiss psychiatrist Carl Gustav Jung agreed that symbols are powerful. He drew a distinction between symbols and signs, defining signs as 'representations of known things' (for example, in the way that a company's trademark represents the company itself). A symbol, on the other hand, is not the 'logical equivalent' of that to which it refers, but 'points beyond itself to an unknown'. In a symbol, the 'known and unknown', or the 'real and unreal', are joined. Jung writes:

> If it were only real, it would not be a symbol, for it would then be a real phenomenon and hence unsymbolic . . . And if it were altogether unreal, it would be mere empty imagining, which, being related to nothing real, would not be a symbol either.[2]

Discussion of Whitehead's symbolism and symbolic reference seems typically to be limited to sense-perception and the use and interpretation of language as symbolic, but Whitehead's position on the two modes of perception insists that should we find 'instances of non-sensuous perception, then the tacit identification of perception with sense-perception must be a fatal error barring the advance of systematic metaphysics' (AI 180). I find, therefore, no convincing argument to limit our discussion of symbolism to human language; on the other hand, since dream symbols are first perceived non-sensuously (directly through the body while the dreamer's conscious brain is inactive), we would be committing that 'fatal error' should we dismiss them from Whiteheadian scholarship.

But how can we connect the dots between Whitehead's 'symbol' and Jung's archetypal images that carry such numinosity and feeling that they appear in dreams and myth, and serve as transformers of psychic energy? Whitehead himself cracks open the door that allows us to connect his ideas on symbolism to the imaginal realm of art, dream symbols and archetypes. In both *Adventures of Ideas* and *Symbolism*, Whitehead extends his definition of symbol to include such instances of creative expression as art, music, play and poetic literature, as well as the symbols of religious ritual, including ceremonial clothing, smells and visual appearances. In discussing Appearance and Reality, he writes that music introduces 'an emotional clothing which changes the dim objective reality into a clear Appearance matching the subjective form provided for its prehension'. Art is said to 'spring' from both physical and 'purely imaginative' origins and to be a 'sublimation' of the 'simple craving to enjoy freely the vividness of life'; it has a 'curative function' and reveals 'intimate, absolute Truth regarding the Nature of Things' (AI 249).

Dreams, according to Jung, are both a source of information and a 'means of self-regulation' (due to the self-balancing nature of the psyche), reveal 'hidden factors of [the dreamer's] personality' and are 'our most effective aids in the task of building up the personality'. Yet even Jung remained mystified, to a degree, about the nature of the dream experience. 'I do not know how dreams arise', he writes. 'But, on the other hand, I know that if we meditate on a dream sufficiently long and thoroughly – if we take it about with us and turn it over and over – something almost always comes of it'.³

Whitehead might very well agree that symbols are both informative and difficult to grasp, as he writes:

> The object of symbolism is the enhancement of the importance of what is symbolized. In a discussion of instances of symbolism, our first difficulty is to discover exactly what is being symbolized. The symbols are specific enough, but it is often extremely difficult to analyse what lies beyond them. (S 62–3)

He describes symbols as having 'unhandy meanings that are often vague' or 'indefinite', but acknowledges as well that 'it is easier to smell incense than to produce certain religious emotions. Indeed, for many purposes, certain aesthetic experiences which are easy to produce make better symbols than do words, written or spoken' (PR 183).

Dream symbols, like linguistic symbols, can be considered a type of 'symbolic truth', wherein the Appearance of the images and the Reality of the dreamer are linked (AI 248). Though there is nothing about a dream image that is directly or causally explicative of the dreamer's reality, upon analysis a dreamer may find the two to be meaningfully related. Of symbolic truths, Whitehead writes, 'In their own natures the Appearances throw no light upon the Realities, nor do the Realities upon the Appearances, except in the experiences of a set of peculiarly conditioned percipients' (AI 248). The dreamer, then, is the 'peculiarly conditioned percipient' whose life experiences and inner psychic reality have conditioned his or her perception and interpretation of dream images.

Therefore, I will argue in this chapter: first, that since Whitehead included imaginal expression in his understanding of symbolism, and was open to including non-sensory perception, we can connect Whitehead's symbolism and that of Jung to broaden and enrich our scholarship; secondly, that functional resonances exist between Whiteheadian and Jungian schools of thought in multiple areas, including the existence of a transpersonal realm of formal causation, a dipolar God or collective unconscious, that feeling and value are integral to human experience, and that raising unconscious material to consciousness is essential for human flourishing; and thirdly, that an integration of Whiteheadian and Jungian thought – especially when combined with a spiritual practice of dream work – can positively influence human society's intensity of experience and our overall aliveness, vitality and zest for life.

Why Jung?

Why introduce a Jungian perspective to a discussion of Whiteheadian thought? Bernard Lee notes that communicating philosophical and theological ideas is difficult because such ideas typically include technical aspects which distance them from the general populace. He sees this as 'regrettable', but not easily avoidable, and writes that if neither Sartre's *Being and Nothingness* nor Heidegger's *Being and Time* could ever be part of the best-seller list, then the chances for Whitehead's *Process and Reality* 'are infinitesimal to the point of practical nonexistence'.[4] Whitehead's philosophy of organism offers much to a world still mired in outdated mechanistic thinking, yet its heavy emphasis on rationality and its intent to appeal to reason might forever relegate it to the margins.

In the early 1990, David Ray Griffin edited a book entitled *Archetypal Process*, a compilation of papers presented at Claremont when depth psychologist James Hillman and other Jungians were invited to dialogue with process thinkers. Griffin notes in his introduction to this text that people are 'moved more by images than by concepts', and reiterates Whitehead's assertion that only 'interesting' propositions can become 'lures for feeling'. Regarding the lack of success of the Whiteheadian movement relative to the Jungian movement, Griffin notes that 'Although we may have rationalized our ineffectiveness by telling ourselves that we were casting pearls before swine, we were actually – to use another New Testament image – giving hungry people stones, in the form of indigestible concepts'.[5] It can be argued that bringing the imaginal perspective inherent in Jungian thought to the table could offer needed balance and persuasiveness.[6]

Whitehead: Purpose, Possibility and Proposition

The process of concretion that every actual entity undergoes is generally described as having three indivisible phases: (1) a 'physical' phase, in which objective data from the actual past are inherited; (2) a 'mental' phase, during which possibilities or potentials are received and evaluated; and (3) an 'integration' phase, during which the entity's physical and conceptual prehensions and feelings are integrated and a decision is made regarding what form(s) the entity will exhibit. In lower-grade organisms, the third phase ends with simple comparative feelings, resulting, by and large, in repeating the past.[7] In order to be a 'concrete fact', actual entities must choose to exhibit some 'determinate form' based on possibilities that exist outside of space and time as 'eternal objects'. Existing as entities, but not actualised, eternal objects 'transcend any occasion in which they are realized',[8] and may be prehended on a purely conceptual basis, or 'impurely' through their presence in past occasions.[9]

In process thought, every actual entity (or 'drop of experience') makes a decision that 'constitutes its own definiteness' in relation to some kind of aim or purpose, 'otherwise there would be random chaos' rather than meaning in existence. If the end was determined strictly by the past, then there would be no actual freedom or true novelty. Rather, the 'end' actually comes from the future as a lure or possibility, and it is God who 'so orders possibility as to render it into a relevant lure for each new experience'.[10] It is this lure – without coercion – that constitutes there even being decision, such that there is more 'order and direction and novelty' in life than could be possible if all were determined by the past or by chance.[11]

The purposeful end or 'aim' that comes as the lure from the future is called the 'initial aim' or 'initial subjective aim'. It is the initial aim that 'makes possible the commencement of process, and which therefore makes reality possible'.[12] This aim is a 'hybrid physical feeling', meaning that it is a physical prehension or ingression of God's conceptual feeling;[13] it is both the entity's 'living immediacy' and its initial standard for experiencing and determining value.[14] For Lee, the initial aim is 'God's purpose at work', and it is through this aim that value, intensity, harmony and beauty enter the world, 'which is to say that creative *advance* is the overall characteristic of process'.[15]

In Whitehead's system, the world 'conspires to produce a new creation' by entering into every new entity. As an activity, an occasion has 'modes of functioning which jointly constitute its process of becoming', and one aspect included in this process is the antecedent object or datum, which provokes 'some special activity of the occasion in question' (AI 176). The object or datum is something that is given, objective; it is not generated within the occasion itself (AI 178–9). Physically, the facts of the world, along with God's 'consequent nature', are prehended or *felt* by the concrescing occasion. Mentally, eternal objects as 'envisaged potential' or forms with subjective form or feeling from God's 'primordial nature' engender novelty by entering into, and being positively prehended by, an actual entity. Within this process, the eternal object has two functions – determining both the datum and the subjective form – and is therefore relational (PR 164).

The integration of what is formal with what is actual first occurs in the mental pole, which has two working phases: (1) 'conceptual reproduction', in which eternal objects are 'forming the data;' and (2) 'conceptual reversion', the phase wherein novelty is 'conceptually felt'. Here, subjective form is 'enriched' and the presence of 'contrast' increases the intensity of the experience (PR 249). Because there are both 'determinateness', which enters each concrescing entity from the actual world in its 'physical inheritance', and 'indeterminateness', from eternal objects in the form of conceptual prehensions, or 'the basic operations of mentality', Whitehead describes the world as 'bipolar' or 'dipolar' (PR 33). It is through this polarity that 'novel determinateness of feeling' enters into the actual world (PR 108). An actual entity is originated through its *physical* side's 'determinate feelings of its actual world', and its *mental* side's 'conceptual appetitions' (PR 45). He writes that:

> The integration of the physical and mental side into a unity of experience is a self-formation which is a process of concrescence, and which by the principle of objective immortality characterizes the creativity which transcends it. So though mentality is non-spatial, mentality is always a reaction from, and integration with, physical experience which is spatial. (PR 108)

It is during this integration phase that consciousness emerges in higher-grade organisms. As mentioned above, in lower-grade organisms the integration of physical and mental prehensions results in a simple comparison; at this point the decision is made and the entity's becoming terminates, while its objective immortality as a 'stubborn fact' for future entities begins.[16] But in higher-grade organisms the integration phase of concrescence is prolonged and propositions are formed that may 'lure' the entity towards possibilities other than mere repetition.[17]

Thomas Hosinski describes propositions as 'lures for feeling formed by integrating an eternal object (form of definiteness) with physical prehensions of actual entities'.[18] Regardless of the truth or falsity of a proposition, its purpose is simply to 'influence the concrescence of actual entities', or to 'lure our action' by 'attracting us through value'.[19] While entertaining propositions, an actual entity unconsciously 'feels' the difference between the 'fact' of past entities and the 'theory' of the new possibilities prehended; this is known as a propositional feeling and is an unconscious valuation. If the integration phase is prolonged even more, such a proposition may become an intellectual feeling, wherein the contrast between 'what is' and 'what might be' is both *felt* and *known* consciously.[20]

The importance of consciousness to higher-level organisms and human society cannot be overstated. It is the consciousness achieved in intellectual feelings that allows the subject 'to criticize [its] propositional lures'.[21] Hosinski notes that:

> conscious intellectual feeling enables the concrescing subject to become consciously aware of at least some aspects of its complex unconscious experience . . . [such intellectual feelings] assist the formation of the occasion's subjective aim . . . [and introduce] the critical ability to form a judgment before [the subject] commits itself to the possibilities contained in the propositional feelings . . . without consciousness, valuations and commitments are 'blind' or unconscious.[22]

Consciousness lays the foundation for reason and rational thought, and brings such gifts as 'increased intensity of experience', and the ability to transcend the self towards loftier ideals and 'a concern for truth, value, [and] quality of human activity and life'.[23] Moreover, 'high grade', conscious occasions bring novelty and originality to life, and symbols play a role in this process, as Whitehead's symbolic reference is an intellectual feeling.[24] It is the knowledge that arises through intellectual feelings that produce the data 'which initiates rational or reflective thought', and it is only 'active thought' that can 'save symbolically conditioned action from quickly relapsing into reflex action' (S 82).

Causal Efficacy and Symbolic Reference

In order to understand symbolic reference, we must briefly examine Whitehead's modes of perception. Though he agreed with David Hume that our sense-perceptions and the things themselves are not one and the same, Whitehead delineated two modes of perception: *presentational immediacy* and *causal efficacy*. Whitehead's world is organically interrelated and enters causally into every new entity in the mode of causal efficacy. Here 'antecedent actual occasions' that are 'causally efficacious' to the 'presented locus' are perceived *directly* through the 'ingression (or "participation")' of influential entities

(PR 169, 180). This mode of perception is active in the initial phase of every concrescing occasion, when the concrescing entity physically prehends the actual facts of its given world of past occasions. Hosinski describes this process through an example of the way in which a physiological feeling of anger is felt prior to sense-perception.[25]

For Whitehead causal efficacy is the foundational mode of perception upon which the mode of presentational immediacy emerges, allowing us to perceive sense-data.[26] The 'interplay' between these two modes of perception is what Whitehead calls 'symbolic reference', and this 'reference' requires there to be 'common ground' between what is being perceived and what is doing the perceiving (PR 168). In Whitehead's system – as in Paul Tillich's theology – the reason a symbol can function as a symbol is because it is both an element of our psyche or consciousness, and yet at the same time it 'participates in the reality to which it gives access'.[27] This *participation* is what Whitehead termed *symbolic reference*.

Symbolic reference is one of the principles that 'govern all symbolism'. It requires that there be two kinds of 'percepta', and that they must be correlated through some kind of 'common ground' (PR 180). Then there is 'symbolic reference' between the two species when the perception of a member of one species evokes its correlate in the other species, and precipitates upon this correlate the fusion of feelings, emotions and derivate actions, which belong to either of the pair of correlates, and which are also enhanced by this correlation. The species from which the symbolic reference starts is called the 'species of symbols' and the species with which it ends is called the 'species of meanings' (PR 181).

Such 'common ground' is not limited to earth-bound actual entities. Whitehead understood God's relationship to the world as one of mutual immanence, and writes that 'It is as true to say that the World is immanent in God, as that God is immanent in the World' (PR 348). In a way that is similarly evocative, Tillich describes the relationship between God and humans in this way:

> In the human spirit's essential relation to the divine Spirit, there is no correlation, but, rather, mutual immanence . . . If God were not also in [humans] so that [humans] could ask for God, God's speaking to [humans] could not be perceived by [humans].[28]

In Whitehead's system, God is immanent in every occasion as its initial subjective aim, in the ordered relevance of possibilities offered as eternal objects, and as God's superjective consequent nature.

Jung: Individuation, Archetypes and Archetypal Images

In Jung's psychology, the psyche 'mediates all experience' and is composed of a personal conscious ego and the personal unconscious. The contents of the personal unconscious remain outside of awareness until they are brought to consciousness through inner work;[29] anything that comes into awareness takes on the quality of consciousness, and what does not remains unconscious. The psyche's 'organ of awareness' is referred to as the 'ego', yet 'a whole other sphere lies outside the ego. This is characterized as the non-ego field, the unconscious.' The unconscious is not concentrated, but 'shades off into obscurity', is 'highly extensive' and 'can juxtapose the most heterogeneous elements in the most paradoxical way'.[30] The interplay between the conscious and unconscious aspects of the psyche is dynamic, and so they are understood as being part of the same system, though not alike.[31]

Due to Jung's repeated observations of patients who – independent of their personal experience or history – related psychic contents to common myth motifs, he concluded that these contents did not originate in the 'space-time world or the individual consciousness', but that there must also be a broader, transpersonal and *collective* unconscious. He was therefore compelled to assume that there were some kind of 'myth-forming' structural elements in the unconscious psyche which 'produced out of themselves revivals of these mythologems, independent of all tradition'. Jung called these structural elements *archetypes*, meaning 'a pre-existent form ... [or] organizing principle in the collective unconscious'.[32] He saw them as 'systems of readiness' that are inherited as a psychic aspect of brain structure.[33]

When he first used the term in 1919, Jung understood the 'archetype' as the 'primordial image' or the instinct's 'self-portrait'. By 1946, Jung began distinguishing between archetypes and 'archetypal images', explaining that the 'archetypal representations (images and ideas) mediated to us by the unconscious should not be confused with the archetypes as such. They are very varied structures which all point back to one essentially "irrepresentable" basic form',[34] in the same way that a staggering number and variety of breeds exhibit the form 'dog'. Clinical psychologist and author Robin Robertson notes that Jung's idea of archetype

> moved increasingly toward abstraction as he gradually stripped away the 'clothing' the archetype wears when it emerges into consciousness ... gradually he came to realize that the [archetypal] image was only

a manifestation of an essentially content-free archetype . . . Finally, late in life, he speculated that perhaps when all personifications were removed from the archetype, we arrive at *number*, which 'may well be the most primitive element of order in the human psyche . . . an *archetype of order* which has become conscious'.[35]

It is critical to understand that Jung saw the collective unconscious both as an objective reality *and* as archetypal imagination ('atemporal envisagement of pure possibility)'.[36] He understood the archetype *itself* as something transcendent or *psychoid* that is *not capable of being made conscious* but which can only present images to the psyche. These images, he surmised, may differ somewhat from the archetype that generated the 'representation'.[37] We can imagine the collective unconscious simultaneously as containing irrepresentable, unconscious and universally available forms and as an objective 'storehouse' of human memory and symbols. Formal archetypes mediate the images and ideas to consciousness, where they interact with the contents of the personal unconscious, especially with what Jung called the 'feeling-toned complexes'.[38] As a result of their movement from the unconscious to the contents of conscious experience, the archetypal images become contextual, numinous and laden with feeling.

Whitehead, Jung and Human Flourishing

When proposing linkages or resonances between Whitehead and Jung, we must keep in mind that each of these men engaged in speculation to one degree or another about the nature of reality, although from differing bases of observation or experience. While Whitehead developed his own terminology to describe the metaphysical workings of the cosmos, Jung developed his terminology to describe the empirical workings of the human psyche. It is therefore not particularly helpful to assert or dispute whether or not there are direct one-to-one correlations between concepts from each school of thought; a more fruitful exercise would be to connect the *functions* each thinker used his terminology to describe.

Both Whitehead and Jung were concerned with common flourishing. Whitehead understood God as the 'organ of novelty, aiming at intensification' (PR 67) and such intensity of experience is a favourable goal for societies. In sections VI and VII of his chapter 'Order of nature' in *Process and Reality*, Whitehead discusses societies of occasions and how they handle environments that are changing their degree of structure and level of complexity. He argues that societies that are less complex will be more stable overall, but will suffer from a lack of intensity of experience. On the other hand, societies that are more complex and more heterogeneous will be less stable and therefore less likely to survive (PR 101).

Ideally, a society should be complex while at the same time able to respond positively to change and to advance in novelty, yet retain enough order and stability to be able to survive and endure over time. But how can a society achieve this? Whitehead posits two ways, both of which depend on 'enhancement of the mental pole, which is a factor in intensity of experience' (PR 101). The first is through 'blocking out unwelcome detail' in the variety of members of a nexus, and the second is by developing one's ability to 'receive the novel elements of the environment into explicit feelings with such subjective forms as conciliate them with the complex experiences proper to members of the structured society' (PR 101-2). Restated in a slightly more 'digestible' way, one can enhance one's mental pole by improving one's ability to discern the initial aims presented by God that offer novel ideas appropriate to oneself, and then integrate and align with those initial aims during one's process of concretion. That member, and that society, will then enjoy more intensity of experience and be more stable as well.

Although Jung wrote almost exclusively about human psychology, he did not truly see humans as separate from the rest of the world and also perceived of a soul of nature.[39] As noted above, Whitehead understood the flourishing of higher-grade organisms to occur through alignment with God's initial aim and through the integrative activities leading to consciousness in intellectual feelings and thereby to rational thought. Moreover, Whitehead described that one could experience 'religious intuitions', which are 'intuitive feelings of the consequent nature of God and God's love for the world'.[40]

For Jung, individuals flourish by being in dynamic relationship with the unconscious, through bringing unconscious material to consciousness and integrating that material over one's lifetime into a whole personality. Unconscious material is brought to consciousness by working with the images that arise in dreams (and through other kinds of inner work); such images are thought to be mediated by the archetypal Self, the God-image in the psyche. The Self is understood as a pattern of wholeness in the psyche, and is both the process of building the personality and its *telos*. According to Ann Belford Ulanov and Alvin Dueck, through the Self we encounter the 'living God' that gives us 'food from the heart of reality'.[41] She notes:

The unconscious communicates through the body, both individually and corporately, through affect-laden images that picture instincts, the thrumming of the mysterious gift of life itself . . . We have pictures that rise up in us autonomously, spontaneously because that is the psyche's language.[42]

This is what Freud called 'primary process thinking' and what Jung called 'non-directed thinking'; both were referring to a primordial level of human experience.[43]

Dipolar God and Collective Unconscious

Grant Maxwell, a scholar in the emerging field of archetypal cosmology, argues that Whitehead and Jung both express a return to formal causation through their functional understanding of their concepts of the nature of God and the collective unconscious, as well as of eternal objects and archetypes.[44] A closer examination indeed reveals resonances between Jung's collective unconscious and Whitehead's dipolar God. Though Gerald H. Slusser writes that God 'may be thought of fruitfully as "the archetype of archetypes"' and that 'this may make contact with Whitehead's notion of God as primordial',[45] a much stronger argument is made by Steve Odin, who equates the unity of God's primordial *and* consequent natures with the collective unconscious. Odin describes both Jung's collective unconscious and Whitehead's dipolar God as a 'synthesis of imagination (the projection of future possibilities) and memory (the restoration of the past in the present)'.[46] Jung did not claim that God and the collective unconscious were identical, but he did assert that God could *only* act upon humans through the psyche, and so it was empirically impossible to distinguish between God and the collective unconscious.[47]

For Jung, archetypes function as an *a priori* pattern that provides a 'cluster of images and associated behavioural and emotional possibilities of response' that then organise the contents of consciousness.[48] The archetypes 'proceed from an unconscious, i.e., objective, reality which behaves at the same time like a subjective one'.[49] The collective unconscious is not acquired through personal experience, but is an inborn 'psychic substrate of a suprapersonal nature' that is common to all humans.[50] He did not reify it, but saw it as an 'archetypal imagination, understood as a dynamic image-making function'.[51]

Jung believed archetypes were closely related to instincts, and recognised the presence of 'patterns of behavior' in all living species.

> Instinct and the archaic mode meet in the biological conception of the 'pattern of behavior.' There are in fact no amorphous instincts, as every instinct bears in itself the pattern of its situation. Always it fulfills an image, and the image has fixed qualities. The instinct of the leaf-cutting ant fulfills the image of ant, tree, leaf, cutting, transport, and the little ant garden of fungi. If any of these conditions is lacking, the instinct does not function, because it cannot exist without its total pattern, without its image. Such an image is an a priori type. It is inborn in the ant prior to any activity for there can be no activity at all unless an instinct of corresponding pattern initiates and makes it possible.[52]

According to Jung, humans have no choice but to act in a 'specifically human way' (as opposed, we may presume, to adopting the instinctual behaviours of spiders or dinosaurs) and the images associated with instinctual patterns of behaviour might be thought of as the meaning of the instinct.[53] In process thought, we might say that such patterns of behaviour or instinct are the result both of possibilities presented in the form of eternal objects and also of the persistent, relevant past that each occasion inherits. Whitehead, too, notes the presence of 'habits' of both behaviour and interpretation functioning among human beings (AI 249). He describes 'modes of functioning' that over time become so important to social groups that 'restless intellectuals' of the group interpret them and raise them from the 'penumbra' of consciousness to a more intellectual level. Such patterns then 'take on the role of an apparatus for expression'. The behaviour patterns themselves (with their entwined emotions) can then evoke the intellectual construction, and Whitehead understands this type of development to be the same as that of religious rituals and ceremonies (AI 249–50). It can be noted as well that the sense that ideas or behaviours can become 'entwined' with emotion evokes Jung's 'feeling-toned complexes'.

Odin argues that Whitehead's primordial and consequent natures of God are performing the same 'cosmological role' as Jung's collective unconscious, and describes the primordial nature as 'the atemporal mental envisagement of all possibilities or archetypal value-patterns, or what are termed "eternal objects" in Whitehead's categoreal scheme'.[54] Odin further notes that occasions are initiated by God through *prescribed* 'archetypal patterns for experiential synthesis'.[55] Similarly, Jung understood the archetypes as pressing 'for their own resolution'[56] and as 'complexes of experience that come upon us like fate'.[57] Elizabeth M. Kraus insists that we should not understand eternal objects as 'inert forms waiting to be appropriated'; rather, they are 'dynamic' and 'have the unrest of the Platonic Eros'.[58]

Robertson also links archetypes to feeling, not just as emotion but as valuation, noting that archetypes have 'feeling connections to things in the world' and are evaluative not based on 'what' something is but in 'how it fits into things'.[59] Walter Shelburne draws from Jung when he writes:

> to a large extent, then, what we add to the picture of archetype by calling the archetypal images 'symbols' is an emphasis on the living intensity of the archetypes as they are experienced. [. . .] 'They are as much feelings as thoughts; . . .' [. . .]. This characteristic quality of the symbol to evoke emotion is termed its 'numinosity', the numen being the specific energy of the archetypes.[60]

This sense that archetypal images or symbols evoke 'feeling' or 'numinosity' is functionally related to what Whitehead calls 'subjective form', and it is subjective form that carries emotion, value and purpose (PR 33, 70). When a feeling has the 'datum' of an eternal object, it is known as a 'conceptual feeling' and it is then a determinant of the character of the concrescing entity (PR 240). 'Feeling', 'emotion' and 'purpose' are important to the process of concretion, and that which is a 'datum for feeling has a unity as *felt*' (PR 24). Whitehead considered feeling to be 'an essential doctrine in the philosophy of organism' because:

> the primary function of a proposition is to be relevant as a lure for feeling . . . The 'subjective aim', which controls the becoming of a subject, is that subject feeling a proposition with the subjective form of purpose to realize it in that process of self-creation. (PR 250)

In Whitehead's philosophy God is 'the lure for feeling, the eternal urge of desire' (PR 344).

An unusual degree of resonance between Whitehead's eternal objects and Jung's archetypes is also apparent when we examine two independent passages about mathematics. In *Science and the Modern World*, Whitehead ponders Pythagoras's question on the 'status of mathematical entities', pointing specifically to the number two and noting that such a number is both 'involved in the real world' *and* 'exempt from the flux of time'. He concludes that such formal elements as number and shape are what lie 'at the base of the real world' (SMW 27–8).

In an eerily similar way, Marie-Louise von Franz, an analyst and student of Jung, writes in *Man and His Symbols* (a text edited by Jung just prior to his death in 1961) that the number two illustrates how numbers 'are not concepts consciously invented by [humans] for purposes of calculation: They are spontaneous and autonomous products of the unconscious – as are other archetypal symbols'. Yet such numbers 'adhere' to objects in the space-time world. She notes:

> Even if we strip outer objects of all such qualities as color, temperature, size, etc., there still remains their 'many-ness' or special multiplicity. Yet these same numbers are also just as indisputably part of our own mental set-up – abstract concepts that we can study without looking at outer objects. Numbers thus appear to be a tangible connection between the spheres of matter and psyche.[61]

Functioning in the mental pole, it could be said that just as the eternal object determines and mediates the datum, the archetype determines and mediates the image. At this stage, what is being described is still objective and unconscious, and is therefore in a state of 'transcendent decision', or is purely conceptual. This datum or image – whose 'relevance' could also be described as an appropriateness for one's personal unconscious and complexes – then 'provokes the origination' of prehension, or the entry into one's personal psyche, where it becomes an 'immanent decision' (PR 164). As an 'immanent decision', there is subjective form, affective tone, and the encounter with one's 'feeling-toned complexes' (AI 76–7). At this stage, the activity is now an 'impure prehension' of an objective datum, image or symbol that the concrescing actuality appropriates for itself (PR 164).

Functioning in the physical pole, archetypal images may be prehended in various types of past occasions, including the dream images generated as signals by the 'complex amplifier' of the human body (PR 182), from historical occasions in the actual world, the 'massive presence of the past',[62] and from God's consequent nature. Archetypal images continue to come into play during the integration phase of concrescence as propositional feelings. Though archetypal dream images are initially perceived through the mode of causal efficacy, in a practice of dream work, they are recorded and analysed, and therefore are then re-perceived through the mode of presentational immediacy. Such dream images brought to conscious awareness are then raised from unconscious propositional feelings to conscious intellectual feelings, where they participate in symbolic reference. Such symbols and images are the linchpin between Whitehead's modes of perception (causal efficacy and presentational immediacy) because they are both present in the psyche and participate in that towards which they point. Clearly, both Whitehead and Jung connect the 'spheres of matter and psyche' through the objective function of the primordial realm of eternal objects and archetypes, as well as in the consequent realm of past fact, propositions, and numinous, historically conditioned imaginal symbols.

Conclusion

If it is true that God is aiming for novelty and intensity of experience, and if these can be realised through enhancing the functioning of the mental pole, aligning with God's initial aim, and bringing more consciousness to our decisions, then it is critical to our experience of aliveness, vitality and zest for life that we find a way to bring discernment and awareness to our moments of concrescence. The questions we must ask now are these: (1) How can we better perceive God's initial aims so as to better conform to each aim presented? (2) How can we raise unconscious propositional feelings to the consciousness of intellectual feelings, and thereby gain the insights needed to make conscious judgements? (3) How can we discern and understand 'religious insights' and 'intuitive judgements' *as such*, so that our experience includes *awareness* of God's consequent nature and love for the world as present in our daily experience? For Jung, and for those drawn to his theories, it is through one's inner life and work with the images that arise from the unconscious that one has access to guidance or direction from that which is greater than the conscious ego, and Jung believed that one of the most effective means of accessing this realm is through the analysis of one's dreams.

If we return now to Whitehead's assertion that 'expression is symbolism' (S 62), then this gorgeous passage from *Religion in the Making* gives us a beautiful sense of its possibilities (I will add the term 'symbolism' wherever Whitehead uses 'expression'):

> Expression [Symbolism] is the *one fundamental sacrament*. It is the outward and visible sign of an inward and spiritual grace ... [without] interpretation, the modes of expression [symbolism] remain accidental, unrationalized happenings of mere sense-experience; but with such interpretation, the recipient extends his apprehension of the ordered universe *by penetrating into the inward nature of the originator of the expression [symbol]*. There is then a *community of intuition* by reason of the sacrament of expression [symbolism] proffered by one

and received by the other. But the expressive [symbolic] sign is more than interpretable. It is creative. It elicits the intuition which interprets it. It cannot elicit what is not there. A note on a tuning fork can elicit a response from a piano. But the piano has already in it the string tuned to the same note. In the same way the expressive [symbolic] sign elicits the existent intuition which would not otherwise emerge into individual distinctiveness. Again in theological language, the sign [Jung might say symbol] works *ex opera operato*, but only within the limitation that the recipient be patient of the creative action. (RM 117–18, emphasis added)

Extending Whitehead's closing metaphor, God is the tuning fork, the human psyche is the piano, the eternal objects/archetypes are the strings tuned to various notes and the archetypal images or symbols are the notes, common to both, in vibration. With their profound capacity for engendering healing in the psyche, it could certainly be said that dreams are a fundamental sacrament, or, as Jung himself said about our experience of dreaming, 'Every night a Eucharist'.[63]

Notes

1 Cirlot, *A Dictionary of Symbols*, x.
2 Quoted in Shelburne, *Mythos and Logos*, 43.
3 Jung, *Modern Man in Search of a Soul*, 62.
4 Lee, *The Becoming of the Church*, 4.
5 Griffin, 'Introduction, archetypal psychology and process philosophy', 15.
6 Though it could certainly be argued that Whiteheadian thought is experiencing a resurgence in certain fields such as ecology, it is less well known than Jungian psychology. On the other hand, it could be argued that the promise of both schools of thought is currently under-exploited.
7 Hosinski, *Stubborn Fact and Creative Advance*, 90.
8 Kraus, *The Metaphysics of Experience*, 40.
9 Kraus, *The Metaphysics of Experience*, 47–8.
10 Cobb Jr, 'Spiritual discernment in a Whiteheadian perspective', 358.
11 Cobb Jr, 'Spiritual discernment in a Whiteheadian perspective', 359.
12 Lee, *The Becoming of the Church*, 94–5.
13 Sherburne, *A Key to Whitehead's Process and Reality*, 49.
14 Hosinski, *Stubborn Fact and Creative Advance*, 173.
15 Lee, *The Becoming of the Church*, 94–5.
16 Kraus, *The Metaphysics of Experience*, passim.
17 Hosinski, *Stubborn Fact and Creative Advance*, 99.
18 Hosinski, *Stubborn Fact and Creative Advance*, 105.
19 Hosinski, *Stubborn Fact and Creative Advance*, 100, 102.

20 Hosinski, *Stubborn Fact and Creative Advance*, 108–13.
21 Hosinski, *Stubborn Fact and Creative Advance*, 115.
22 Hosinski, *Stubborn Fact and Creative Advance*, 113.
23 Hosinski, *Stubborn Fact and Creative Advance*, 124.
24 Hosinski, *Stubborn Fact and Creative Advance*, 118.
25 Hosinski, *Stubborn Fact and Creative Advance*, 68.
26 Lee, *The Becoming of the Church*, 102.
27 Quoted in Lee, *The Becoming of the Church*, 105.
28 Tillich, *Systematic Theology, Volume III*, 114, 127.
29 Singer, *Boundaries of the Soul*, 15.
30 Jung, *Modern Man in Search of a Soul*, 186.
31 Singer, *Boundaries of the Soul*, 15.
32 Singer, *Boundaries of the Soul*, 100.
33 Card, 'The archetypal view of C. G. Jung and Wolfgang Pauli', 23.
34 Jung, *The Basic Writings of C. G. Jung*, 105–6.
35 Robertson, 'The evolution of Jung's archetypal reality', 66.
36 Odin, *Process Metaphysics and Hua-Yen Buddhism*, 172.
37 Jung, *The Basic Writings of C. G. Jung*, 106.
38 Jung, *The Basic Writings of C. G. Jung*, 359.
39 For more on this topic, see Sabini, *The Earth Has a Soul*, as well as Jung's individual work and collaborative work with Wolfgang Pauli on synchronicity.
40 Hosinski, *Stubborn Fact and Creative Advance*, 201–2.
41 Ulanov and Dueck, *The Living God and Our Living Psyche*, 69.
42 Ulanov and Dueck, *The Living God and Our Living Psyche*, 72.
43 Ulanov and Ulanov, *Religion and the Unconscious*, passim.
44 Maxwell, 'Archetype and eternal object', 51–67.
45 Slusser, 'Jung and Whitehead on self and divine', 90.
46 Odin, *Process Metaphysics and Hua-Yen Buddhism*, 160.
47 Odin, *Process Metaphysics and Hua-Yen Buddhism*, 165, 167.
48 Ulanov, *The Functioning Transcendent*, 98.
49 Jung, *The Basic Writings of C. G. Jung*, 127.
50 Jung, *The Basic Writings of C. G. Jung*, 359.
51 Odin, *Process Metaphysics and Hua-Yen Buddhism*, 171.
52 Jung, *The Basic Writings of C. G. Jung*, 90.
53 Jung, *The Basic Writings of C. G. Jung*, 90.
54 Odin, *Process Metaphysics and Hua-Yen Buddhism*, 160.
55 Odin, *Process Metaphysics and Hua-Yen Buddhism*, 168.
56 Jung, *The Basic Writings of C. G. Jung*, 193.
57 Jung, *The Basic Writings of C. G. Jung*, 391.
58 Kraus, *The Metaphysics of Experience*, 47.
59 Personal phone conversation with Robin Robertson held in the autumn of 2015.
60 Shelburne, *Mythos and Logos*, 43–4.

61 von Franz, 'Conclusion: science and the unconscious', 310.
62 Sherburne, *A Key to Whitehead's Process and Reality*, 113.
63 Hudson, *Natural Spirituality*, 81.

Bibliography

Card, Charles R., 'The archetypal view of C. G. Jung and Wolfgang Pauli', *Psychological Perspectives*, 24 (spring–summer 1991), 23–33.

Cirlot, Juan Eduardo, *A Dictionary of Symbols*, second edition (New York: Barnes and Noble Books, 1993).

Cobb Jr, John B., 'Spiritual discernment in a Whiteheadian perspective', in Harry J. Cargas and Bernard Lee (eds), *Religious Experience and Process Theology: The Pastoral Implications of a Major Modern Movement* (New York: Paulist Press, 1976).

Griffin, David Ray, 'Introduction, archetypal psychology and process philosophy: complementary postmodern movements', in David Ray Griffin (ed.), *Archetypal Process: Self and Divine and Whitehead, Jung, and Hillman*, first edition (Evanston: Northwestern University Press, 1990).

Hosinski, Thomas, *Stubborn Fact and Creative Advance* (Lanham: Rowman and Littlefield, 1993).

Hudson, Joyce Rockwood, *Natural Spirituality: Recovering the Wisdom Tradition in Christianity*, second edition (Danielsville: JRH Publications, 2001).

Jung, C. G., *Modern Man in Search of a Soul* (Orlando: Harcourt Harvest, 1955).

Jung, C. G., *The Basic Writings of C. G. Jung*, reprint edition, ed. Violet Staub De Laszlo (New York: Modern Library, 1993).

Kraus, Elizabeth M., *The Metaphysics of Experience: A Companion to Whitehead's Process and Reality* (New York: Fordham University Press, 1979).

Lee, Bernard J., *The Becoming of the Church: A Process Theology of the Structures of Christian Experience* (New York: Paulist Press, 1974).

Maxwell, Grant, 'Archetype and eternal object: Jung, Whitehead, and the return of formal causation', *Archai: The Journal of Archetypal Cosmology*, 3 (winter 2011), 51–71.

Odin, Steve, *Process Metaphysics and Hua-Yen Buddhism: A Critical Study of Cumulative Penetration vs. Interpenetration* (Albany: State University of New York Press, 1982).

Robertson, Robin, 'The evolution of Jung's archetypal reality', *Psychological Perspectives* 41:1 (17 January 2008), 66–80.

Sabini, Meredith (ed.), *The Earth Has a Soul: C. G. Jung's Writings on Nature, Technology and Modern Life* (Berkeley: North Atlantic Books, 2002).

Shelburne, Walter, *Mythos and Logos in the Thought of Carl Jung: The Theory of the Collective Unconscious in Scientific Perspective* (Albany: State University of New York Press, 1988).

Sherburne, Donald W. (ed.), *A Key to Whitehead's Process and Reality* (Chicago: University of Chicago Press, 1966).

Singer, June K., *Boundaries of the Soul: The Practice of Jung's Psychology*, revised edition (New York: Anchor Books, 1994).

Slusser, Gerald H., 'Jung and Whitehead on self and divine: the necessity for symbol and myth', in David Ray Griffin (ed.), *Archetypal Process: Self and Divine and Whitehead, Jung, and Hillman*, first edition (Evanston: Northwestern University Press, 1990).

Tillich, Paul, *Systematic Theology, Volume III* (Chicago: University of Chicago Press, 1976).

Ulanov, Ann Belford, *The Functioning Transcendent*, first edition (Wilmette: Chiron Publications, 1996).

Ulanov, Ann Belford, and Alvin Dueck, *The Living God and Our Living Psyche: What Christians Can Learn from Carl Jung* (Grand Rapids: Wm. B. Eerdmans, 2008).

Ulanov, Ann Belford, and Barry Ulanov, *Religion and the Unconscious*, new edition (Westminster: John Knox Press, 1985).

von Franz, Marie-Louise, 'Conclusion: science and the unconscious', in C. G. Jung (ed), *Man and His Symbols* (Garden City: Doubleday, 1969).

Watts, Fraser N., *Theology and Psychology* (Farnham: Ashgate Publishing, 2002).

Whitehead, Alfred North, *Adventures of Ideas* (New York: The Free Press, [1933] 1967).

Whitehead, Alfred North, *Process and Reality: An Essay in Cosmology*, corrected edition, ed. David Ray Griffin and Donald W. Sherburne (New York: The Free Press, [1929] 1978).

Whitehead, Alfred North, *Religion in the Making* (Cambridge: Cambridge University Press, [1926] 2011).

Whitehead, Alfred North, *Science and the Modern World* (New York: The Free Press, [1925] 1967).

Whitehead, Alfred North, *Symbolism: Its Meaning and Effect* (New York: Fordham University Press, [1927] 1985).

Part III
Misplaced Concreteness in Ethics and Science

8

A Dog's Life: Thought, Symbols and Concepts

JEFFREY BELL

Do animals have thoughts? What would it mean to say that they do? To these questions, among others, philosophers have often answered that animals do not have thoughts, and if they did, it is assumed, they would speak a language. Aristotle, for example, argues in the *Politics* that whereas the bees are a gregarious, social animal, their lack of speech leaves them unprepared to 'set forth the expedient and inexpedient, and therefore likewise the just and the unjust', and hence it is our ability to think and state what is right and best that makes it possible to live together as 'a family and a state'.[1] Descartes is even more explicit in relegating thought only to those who speak a language – to wit, humans – for 'Language is the only certain sign of thought hidden in a body'.[2] Yet Descartes leaves the door open to the possibility that animals might think, for while language is the 'only certain sign of thought', he admits that we cannot be certain that they do not have thoughts – 'the human mind does not reach into their hearts'.[3] Philosophers in the twentieth century pushed their way through the door Descartes left ajar and began to explore the possibility of non-linguistic thought. Whitehead's extended essay, *Symbolism: Its Meaning and Effect*, was one of the first efforts to do just this.

In this chapter I will connect Whitehead's theory of symbols to recent work on animal thought, especially the work of Elisabeth Camp, in order to begin sketching a theory of the relationship between life and thought. In the case of Whitehead this relationship is particularly important, for a strong case can be made that Whitehead adopts a strong vitalist perspective (what we might call 'panvitalism') in his attribution of the fundamental, basic characteristics of life to all things. Evidence for this viewpoint is easily found in Whitehead's work. In his symbolism essay, for instance, Whitehead says the following: 'A rock is nothing else than a society of molecules, indulging in every species of activity open to molecules' (S 64), and 'All physical response on the part of inorganic matter to its environment is thus properly to be termed instinct' (S 78). If life is at its base instinctual, and if other forms of life are built upon these foundations, as Whitehead argues, then it appears that Whitehead does indeed endorse the view that life is a fundamental attribute of all things, or that he is a panvitalist.

It is on the basis of a panvitalist conception of reality that Whitehead will account for the emergence of thought, and in particular the emergence of the use of symbols. In doing so, however, a case can equally be made that Whitehead is a panpsychist, or that he follows in the footsteps of Plotinus and argues that all reality involves a form of mental contemplation and the individuating unity of a subjective experience.[4] Whitehead is quite forthright in his assertion that 'the life-history of an enduring organism holds for all types of organisms, which have attained to unity of experience, *for electrons as well as for men*' (S 28, emphasis added). Whitehead's theory of symbols, I will argue, is an attempt to delineate the relationship between these two types of process – the panvitalist processes of life and the panpsychic processes of contemplation. It is in this context where the arguments regarding animal thoughts become most relevant, and to that end section I will explore Elisabeth Camp's arguments for the claim that animals such as dogs are indeed capable of what she calls a 'minimalist' form of thought. I will also show how Camp's arguments are to a large extent anticipated by Whitehead in his extended essay *Symbolism: Its Meaning and Effect* (hereafter, *Symbolism*).

In section II I will highlight the limitations of animal thought, limitations that have traditionally led philosophers to turn to the use of language as the key feature of human thought. What is crucial to language, however, is the fact that it is stimulus-independent – that is, through language one can bring to mind a representation of an object or event even in the absence of this object or event. Camp will argue, however – and this argument can be found in Whitehead's *Symbolism* as well – that the development of instrumental skills among animals presupposes a thought that is minimally stimulus-independent. This claim will in fact be one of the central arguments of Whitehead's essay. What language adds to the mix is a fundamental shift in the ability to recombine representations in the absence of stimuli.

Understood in this way, however, language and thought are quite limited, and in section III I will argue that what is critical to thought is the process of deterritorialisation/reterritorialisation. With Camp's pragmatic, instrumental understanding of the foundations of thought,

thought itself becomes tied to a limited set of presupposed determinate ends for the sake of which thought and language functions. Although Camp will recognise that language does provide us the opportunity to entertain absurd thoughts, or to provoke the incredulous stare from our peers, this aspect of thought is only briefly touched upon. In contrast, I will argue, along the lines Whitehead provides for us, that it is thought as de/reterritorialising that characterises the interface between the panvitalist and pansychist tendencies of reality.[5] As Whitehead might state our conclusion regarding thought, it consists of nothing less than events or actual occasions.

I

If we are to begin to argue that animals think, that they make use of concepts and therefore have thoughts that they nonetheless do not express in language, then what might such thinking look like? What evidence would count for such thoughts if we do not have Descartes' 'certain sign of thought' in language? Recent work in ethology and cognitive psychology has pointed, as evidence of their ability to use concepts, to the ability of animals to represent and respond to varying aspects of their environment in a systematic manner.[6] In her essay 'Putting thoughts to work', Elisabeth Camp builds upon this recent work by arguing that while the scientific tradition in ethology provides us with important insights 'about the underlying mechanisms that enable us to think', we are nonetheless in need of the insights of the philosophical tradition, which 'captures an important insight about what thinkers can do with their thoughts'.[7] In her essay, Camp combines the two traditions to provide what she takes to be the necessary criterion for thought – namely thought involves a 'stimulus-independence' that 'results from systematically recombinable, stimulus-independent representational abilities [that are] flexible, abstract, and actively self-generated, and thus quite unlike mere passive reaction to stimuli'.[8]

To begin fleshing out this criterion for thought, let us begin with a dog. I have a Springer Spaniel named Maddie. She is a sweet, nervous and, I must confess, not very well trained or disciplined dog. Maddie is nonetheless very present to many aspects of her environment. When we turn on the downdraft to our cooktop, she flees to the other room out of fear of the noise it makes; when we say 'cheese', she comes running, hoping for a 'treat'; when we say 'chair', she runs to her chair and sits; and when I take her on a walk, she wants to play with the neighbour's Cocker Spaniel, barks in anger at the Labrador across the street and ignores the barking poodle down the street. In all of these cases and many others I could list, Maddie exhibits what Camp refers to as a 'recombinable representational ability'.[9] As Camp clarifies, 'a dog D might encounter several different dogs M, N and O upon multiple occasions and treat each of the dogs differently when it does'.[10] On one occasion, for instance, M's behaviour 'might cause D to treat it as a hunting partner', and yet on another occasion 'D might treat M as a threat'.[11] What the dog is able to do, in other words, is to systematically represent different dogs M, N and O and to recombine these representations with the representations of hunter, playmate and threat. On one occasion, for example, Maddie will identify (i.e., represent) the Cocker Spaniel next door as a playmate and behave as one would expect in this circumstance, but given a different circumstance Maddie can, and has, represented the neighbour's dog as a threat and responded accordingly. A dog, therefore, and to a minimal extent, is capable of exhibiting a key aspect of conceptual thought that Gareth Evans has famously identified as the 'generality constraint'.

Concepts, Evans claims, are fundamentally structured, in that if one has the ability to entertain the thought that 'a is F', Evans claims, then this presupposes that one 'must have the conceptual resources for entertaining the thought that a is G'.[12] In other words, if I can entertain the thought that Peter is sad, then I should also be able to entertain the thought that Peter is happy, or that John, Fred and Sally are happy. Conceptual thought, therefore, presupposes an ability to recombine representations and to generalise from one particular circumstance to another. Maddie does appear able to exhibit this ability to the extent that she responds consistently and systematically to threats and playmates, and so on, but does so in a manner that combines such representations with different dogs – sometimes the neighbour's dog is a threat, sometimes a playmate. This is precisely the point Camp stresses as she extends recent work in ethology. There are limitations to this form of thought, however, and we have not yet attained a thought that is stimulus-independent, that is, a thought that is not prompted by the immediate presence of a stimulus. Maddie, for instance, will not entertain the thought that the neighbour's dog is a threat in the absence of this dog. It is for this reason that Camp claims that while dogs and other animals may employ conceptual representations in a manner in line with Evans' 'generality constraint', they do so minimally and within the limitations of actual stimuli. It is at this point where Camp turns to the traditional intellectualist arguments in defence of language as the 'certain sign of thought', for it is frequently argued that language provides us with the resources necessary to employ thoughts that represent situations and states of affairs in the absence of those very situations and states of affairs.

Before turning to Camp's arguments regarding the intellectualist arguments for thought, let us bring Whitehead into the mix. In his *Symbolism* essay, Whitehead argues that the capacity to think is an ability shared by non-humans. Early in the essay Whitehead refers to the fact that upon seeing a chair we readily pass, especially if tired, 'straight from the perception of the coloured shape to the enjoyment of the chair, in some way of use, or of emotion, or of thought' (S 3). The importance of this example is twofold for Whitehead. On the one hand, it betrays for Whitehead the fact that the transition from 'the perception of the coloured shape' before me to the conclusion that there is a chair before me is not an inference that requires a 'high-grade character of the mentality' (S 3). It is not a great, difficult leap of logical inference, and yet it *is* an inference, and to avoid making this inference is not easily done. In fact, Whitehead claims that his 'friend the artist, who kept himself to the contemplation of colour, shape and position' alone was only able to avoid the inferential transition from colour and shape to chair because he was a 'highly trained man' (S 3). This brings us to the second key lesson Whitehead draws from the example – namely, that the inferential relation associated with perceiving a chair given the perception of the 'colour, shape and position' is an inference of thought that one finds in non-humans, such as dogs. As Whitehead puts it, 'the transition from a coloured shape to the notion of an object which can be used for all sorts of purposes which have nothing to do with colour, seems to be a very natural one; and we – men and puppy dogs – require careful training if we are to refrain from acting upon it' (S 4).

But is this how it works? Do we have an incoming perceptual content that serves as the basis upon which we make our inferential conclusions regarding what is present before use? Are we first given 'colour, shape and position' and then infer from these givens the presence of a chair? In short, is Whitehead beholden to what Wilfrid Sellars calls the 'Myth of the Given' and hence susceptible to Sellars' critique of this myth, a critique that has become influential in the work of John McDowell, Robert Brandom and others?[13] On the one hand, Whitehead's theory of symbols does draw heavily from the distinction he makes between what he calls 'presentational immediacy' and 'causal efficacy' (S 17). That which is given in presentational immediacy 'expresses', Whitehead claims, 'how contemporary events are relevant to each other, and yet preserve mutual independence' (S 16), and causal efficacy introduces 'components which are again analyzable into actual things of the actual world and into abstract attributes, qualities, and relations, which express how those other actual things contribute themselves as components to our individual experience' (S 17). Stated simply, everything given in presentational immediacy presupposes a causal efficacy that accounts for its very givenness. Whitehead is very clear and adamant that what is given in presentational immediacy is not a 'simple occurrence' – an isolated, individuated fact or impression from which we then deduce and arrive at, through an association and connection of these simple occurrences, a belief about what is there before us. Both Hume and Kant, Whitehead argues, accept the 'naïve presupposition of "simple occurrence" for the mere data' (S 38). What is given, therefore, is not an isolated fact or impression; to the contrary (and with this move Whitehead circumvents a Sellarsian-style criticism), what is given are 'actual things of the actual world', and as for actual things, Whitehead argues that 'every actual thing is something by reason of its activity; whereby its nature consists in its relevance to other things, and its individuality consists in its synthesis of other things so far as they are relevant to it' (S 26).

Every actual thing, in other words, is what it is only insofar as it actively constitutes a synthesis of other actual things, and every actual thing is in turn actual only to the extent that it is taken up in a process or activity of being relevant to the synthesis of other actual things.[14] Whatever is given in presentational immediacy, therefore, is not a simple occurrence, a simple given, but is always, for Whitehead, a synthesis and 'unity of experience, for electrons as well as for men' (S 28). It was this claim, mentioned earlier, that led us to point to the panpsychist tendencies in Whitehead's philosophy. We will return to this tendency below, but for the moment what is important to note is that presentational immediacy is part of a process that entails the retention and synthesis of other actual things, a retention and synthesis that produce habits and expectations of occasions to come that will conform to these syntheses. 'A dog', Whitehead claims, 'anticipates the conformation of the immediate future to his present activity with the same certainty as a human being . . . [a] dog never acts as though the immediate future were irrelevant to the present' (S 42). It is for this reason that Whitehead claims that 'presentational immediacy is an important factor in the experience of only a few high-grade organisms' (S 23). In other words, a high-grade organism such as a dog has a variety of actual things from which to synthesise that which is given in presentational immediacy, and it is for this reason that a dog can be mistaken, such as Aesop's dog (to follow Whitehead's example) who dropped the meat he was carrying in his mouth in order to go after the meat he saw reflected in the water, the meat being, of course, the very meat that was in his mouth. What happened in the case of Aesop's dog is that the presentational immediacy of the meat was taken to be other than the meat already present in the dog's mouth – that is, the dog mistook the cause for the presentational immediacy, and it is precisely the synthesis of presentational immediacy and causal efficacy that is for Whitehead the very essence of symbolic reference: 'The synthetic activity whereby these two modes [i.e. presentational immediacy and causal efficacy] are fused into one perception is what I have called "symbolic reference"' (S 18).

As was mentioned earlier, however, Whitehead goes much farther than simply asserting the possibility of symbolic reference and presentational immediacy among 'only a few high grade organisms'. Whitehead argues that electrons, as well as men, attain a 'unity of experience' and have a 'life-history' (S 28). Later in the *Symbolism* essay, Whitehead explicitly and perhaps surprisingly argues that 'A rock is nothing else than a society of molecules', and he does this to 'draw attention to this lowly form of society in order to dispel the notion that social life is a peculiarity of the higher organisms' (S 64). Although only a few high-grade organisms may be capable of symbolic reference, forms of society are apparently shared by all actual things, by anything that is a synthesis of other actual things, which for Whitehead is all things. As Whitehead stated this point in *Process and Reality*, there is nothing but actual occasions (or 'actual things', as they are termed in the *Symbolism* essay). For Whitehead, therefore, to understand thought and symbolic reference we need to place the development of thought in the context of social forms and that which is necessary for the maintenance of social cohesion. We have already seen that Whitehead accepts the notion that thought is an ability humans and a few high-grade organisms such as dogs share, but to gain a better understanding of symbols and our capacity for mistakes, as well as the role of language in thought, we need to turn to a discussion of the relationship between thought and the panvitalism that we claim is an important theme in Whitehead's thought.

II

Earlier we pointed to Whitehead's claims that 'A rock is nothing else than a society of molecules, indulging in every species of activity open to molecules' (S 64), and the related claim that 'All physical response on the part of inorganic matter to its environment is thus properly to be termed instinct' (S 78). It was argued that this is evidence for Whitehead's panvitalism, and we are now in a better position to flesh out the implications of this claim. A rock and other instances of inorganic matter are understood by Whitehead to respond to their environment in the form of instinct in the sense that there are 'environmental obligations' necessary for the survival and continued persistence of the 'society of molecules' that constitute a rock (S 65). There are thus normative constraints that instinct enables the rock to satisfy – namely, a rock need not select what is relevant for its continued survival, for such selections are done automatically through the continued persistence of the life history. The automatic retention and continuation of a past life history are the instinct of substance, both organic and inorganic, though inorganic substances such as a rocks have a hands-down advantage over organic matter in this respect. As Whitehead puts it, 'So far as survival value is concerned, a piece of rock, with its past history of some eight hundred million years, far outstrips the short span attained by any nation' (S 64–5). In other words, a rock is nothing less than the automatic realisation of the norms of its environmental obligations. A rock can do no wrong, for it automatically selects that which is relevant based on its life history, and it is this history that predetermines the continued persistence of the society of molecules that constitute the rock.

As we move on to living beings we move, Whitehead argues, to a form of life that is not purely instinctual (in Whitehead's sense of the term), but rather to life 'conceived as a bid for freedom on the part of the organism, a bid for a certain independence of individuality with self-interests and activities not to be construed purely in terms of environmental obligations' (S 65). A rock's survival is not dependent upon selecting from its environment that which is relevant to its survival. A rock simply is the fulfilment of its environmental obligations. A tick, by contrast, and to draw from Jakob von Uexküll's famous example, lives a life where the society of molecules and occasions that constitute the tick require that it make certain selections within its environment – what Uexküll calls the *umwelt* – in order to survive. It is at this point where the organism, in this case the tick, becomes opened to 'presentational immediacy'. As we saw earlier, whatever is given in presentational immediacy is not a simple occurrence, and hence not a simple given, but is always, for Whitehead, a synthesis and 'unity of experience, for electrons as well as for men' (S 28). This unity is precisely the unity of retentions and expectations that constitute the 'unity of experience'. The unity of experience of a rock is simply its lived history, and this unity does not involve any expectations that will signal some occasions as more relevant to its future than others. Whitehead's panpsychism, therefore, is one that involves a continuum: on one end the unity of experience is purely instinctual, and on the other it involves an active engagement with and selection of that which is relevant to maintaining the unity of experience. A dog is well along on this continuum. Recall Whitehead's point that a 'dog never acts as though the immediate future were irrelevant to the present' (S 42). A tick is closer to the instinctual side of the continuum than a dog, but even for the tick there are three occasions that are relevant to its survival, and these signal behavioural changes that set the tick off on a path to act within the parameters of the normative obligations. Uexküll describes the case of the tick:

> The approach of her prey becomes apparent to this blind and deaf bandit only through her sense of smell. The odour of butyric acid, which emanates from the sebaceous follicles of all mammals, works on the tick as a signal that causes her to abandon her post (on top of the blade of grass/bush) and fall blindly downward toward her prey. If she is fortunate enough to fall on something warm (which she perceives by means of an organ sensible to a precise temperature) then she has attained her prey, the warm-blooded animal, and thereafter needs only the help of her sense of touch to find the least hairy spot possible and embed herself up to her head in the cutaneous tissue of her prey. She can now slowly suck up a stream of warm blood.[15]

The tick is thus signalled to act by three occasions that are relevant to its survival – the odour of butyric acid, the warmth of the animal and the animal's blood, and the feel of hair and skin. Given the appropriate signals, experiments have shown that the tick will automatically respond in the relevant way regardless of whether the signal was indeed related to the appropriate cause – that is, a warm-blooded animal.[16] Ticks can get it wrong. A dog can also get it wrong – as Whitehead's use of the example of Aesop's dog shows – but dogs are also capable of being attuned to many more signs than the tick, and thus the task of determining the relevance of occasions to its immediate future becomes a more complicated process for the dog. As Camp shows, one important complicating factor for dogs is the ability dogs have to recombine representations and to modify representations as a result of experience. A dog may have come to associate a particular dog with being a hunting or playing partner, but a change in this other dog's behaviour may lead to a modification so that it is now represented as an enemy dog. A former owner turned abuser may lead to a similar recombination of representations on the part of a dog, and to very different behaviours that these representations signal (from movement towards to retreat from). Since ticks are not capable of a recombination of representations or the modification of representations as a result of experience, their behaviour is instinctual. Whitehead is keenly attuned to this aspect of animal behaviour. As he argues, the 'more developed type of living communities requires the successful emergence of sense-perception to delineate successfully causal efficacy in the external environment' (S 82). In short, the more an organism requires the successful recognition of signs as signals for occasions relevant to its continued existence, the more presentational immediacy combines with causal efficacy, or the more the organism becomes dependent upon symbolic reference. And symbolic reference, as with instrumental reasoning, involves an element of stimulus-independence in that the sign given in presentational immediacy signals an occasion that is independent of that which is given. The capacity to discern the relevance of signs is integral to animal life, Whitehead argues, and dogs, as we have seen, are well along the continuum of being able to recognise, recombine and act upon the discernment of a wide variety of signs. Despite the added complexities of thought associated with dogs as compared with ticks, for example, most theorists, including Camp, would point out that their thought is prompted by stimuli and would not arise in the absence of stimuli, and thus they are *not* capable of a thought that is stimulus-independent, and it is this that is widely held to be one of the hallmarks of fully fledged thought.

How does thought become stimulus-independent? Traditionally, it is argued that thought is only stimulus-independent among animals that possess a language. As Descartes said, it is the 'only certain sign of thought'. Camp criticises this assumption and argues that instrumental reasoning among animals such as primates provides evidence for thought that is stimulus-independent. Camp in fact argues that 'it should be uncontroversial that the sorts of behaviours that have already been observed require a degree of stimulus-independence'.[17]

In adopting a tool for a specific purpose, for instance, a chimpanzee is attentive to working on the tool in the absence of the location to which the tool will be put to work. There are many other examples along these lines, and thus Camp comes to the conclusion she takes to be uncontroversial, and yet Camp asks whether this 'capacity for instrumental reasoning could provide enough stimulus-independence to underwrite fully systematic recombinability'.[18] Camp offers a direct answer to this question – 'I think the answer is almost certainly not' – and her reasoning is that she doubts that 'a practical ability to reason instrumentally could provide a creature with the ability to entertain all the possible combinations of its representational abilities, whatever its current circumstances'.[19]

The implications of this claim are significant. As Camp goes on to provide further details for her doubts regarding the degree to which non-linguistic animals can 'entertain all possible combinations of its representational abilities', she points to the fact that these animals cannot entertain absurd thoughts. A bee, for instance, as was shown in a series of famous experiments,[20] will not recognise as relevant (and hence as meaningful) a bee dance that would have it fly to the middle of a lake and a chimpanzee will not respond to the prompt to think about a dominant banana. There is certainly no reason for bees and chimpanzees to entertain these thoughts – there is no practical context where a a good source of nectar is in the middle of a lake or a banana is dominant. Do the bees and chimpanzees recognise the absurdity of these thoughts or do they simply fail to recognise them as signalling anything – in short, are they simply irrelevant? In the case of bees Camp admits that 'we lack any possible empirical means for distinguishing active disbelief from either disinterest or blank incomprehension'.[21] Chimpanzees are in a different situation, however, for they can, Camp argues, 'disengage their representational abilities sufficiently from their current situation to understand how to make some *prima facie* irrelevant thoughts – thoughts they apparently have no reason to think – relevant'.[22] The expanded capacity of instrumental reasoning for the chimpanzee enables it to think thoughts that are on the surface irrelevant in an effort to make them relevant. But chimpanzees are still limited. More precisely, Camp argues that for chimpanzees a thought, even a seemingly irrelevant or absurd thought, remains meaningless and empty as long as 'these thoughts have no possible use for them'.[23] Without a practical application for thoughts, Camp argues that 'the bare difference between truth and falsity itself lacks any relevance . . . [and therefore it is unlikely that] the chimpanzees could get enough of an independent grip on either the thoughts themselves or on their truth-values to appropriately assent to and dissent from the corresponding "sentences"'.[24]

It is at this point that the importance of language becomes clear, for a language, Camp notes, 'wears its own recombinability on its syntactic sleeve', and hence the potential to recombine thoughts becomes all the more obvious and it becomes 'easier for a thinker to entertain the full range of its potential conceptual combinations'.[25] Stated in another way, the linguistic manifestation of thoughts allows for a greater network and combination of thoughts precisely because the syntactical nature of language structures it such that the meaningfulness of the thoughts expressed is conditioned by the networked relationships between components. As thoughts emerge within this networked structure, it becomes easier to pursue and develop them through their networked affiliation to other thoughts. It is the linguistic expression of thoughts and the syntactic recombinability they presuppose that makes it possible to entertain thoughts regardless of whether they have practical relevance or not. As the linguist Émile Benveniste has argued, what differentiates human linguistic communication from other forms of animal communication and uses of symbolic reference is that humans are capable of indirect communication.[26] A thought can be communicated solely by being related to another thought rather than being tied, as with the case of bees and chimpanzees, to a situation that has practical relevance. It is this capacity for indirect communication, made possible by the syntactic, recombinable structure of language, that makes it possible to think about the nature of thoughts themselves, to discern their relevance, their truth and falsity. The limitations regarding what can or cannot be thought remain, however, for, as Camp acknowledges:

> We cannot demarcate in advance the boundary between substantive thought and empty nonsense, because the line shifts as our imaginative and scientific horizons expand: the thoughts that the mind is the brain and that matter is energy, for instance, once seemed like cross-categorical nonsense but are now core tenets of established scientific theories.[27]

This running up against the limits of thought is key, but Camp does not fully explore its implications, and in fact seems somewhat resigned to the inevitability of these limits. In the next section we will argue that these encounters with the limits of thought tell us a tremendous amount about the nature of thought.

III

In the final few pages of *Symbolism*, Whitehead makes some fascinating observations about the use of symbols in social and political contexts, and this provides, I will argue, an essential supplement to the understanding of language we have set forth so far in comparing the work of Whitehead and Camp. In particular, what is largely absent from the analysis of the first two sections of this essay is a detailed discussion of the social and political conditions of language and thought. We did begin with Whitehead's panvitalist claim that 'a rock is nothing else than a society of molecules', and thus with Whitehead's emphasis upon the normative obligations essential to the continued *social* existence of all individuals, individuals being nothing less than societies of actual occasions. From there, however, our discussion quickly turned to the capacities of an individual organism to utilise symbolic references in order to discern the signs that signal the relevant actual occasions critical to the continued existence of the individual (recall the tick). Our turn to Elisabeth Camp's work allowed us to extend this discussion to instrumental reasoning as a form of stimulus-independent thinking. In this context the social emerged as one of the important contexts essential for putting thoughts to work in cooperative contexts, but the social and political was not itself taken to be essential to the nature of thought itself.

For Whitehead the social nature of thought is essential. Earlier we discussed Whitehead's claim that all actual things are what they are by reason of their activity and the relevance of this activity to other things, and the individuality of each thing 'consists in its synthesis of other things so far as they are relevant to it' (S 26). Every thing is thus social by nature, and it should thus be unsurprising that thoughts themselves will exemplify a similar structure. This is why the closing pages of Whitehead's *Symbolism* are so important. On the surface, Whitehead's concern at the end of his essay is the extent to which symbolic reference loses its significance and becomes instead a meaningless, quasi-instinctual reaction. Unlike a pure instinctive action – which can be analysed solely 'in terms of those conditions laid upon its development by the settled facts of its external environment', such as the butyric acid and so on that signal the tick to fall from its perch – Whitehead argues that an action founded upon symbolic reference entails a 'perceptive mode of presentational immediacy' (S 78). That is, *the manner in which something is presented* will play an important role in determining the relevance and hence symbolic reference of that which is given in presentational immediacy. The fact that the morning star was presented in a contextual mode that was different from that in which the evening star was presented led many, as Frege famously pointed out, to assume that the symbolic referent, or the causal efficacy of these presentations, was different in each case. This leads Whitehead to conclude that there 'is no sense in which pure instinct can be wrong. But symbolically conditioned action can be wrong, in the sense that it may arise from a false symbolic analysis of causal efficacy' (S 81). To contrast the sense in which instinct cannot be wrong from that in which a symbolically conditioned action can be, we can return to the tick. Earlier we said the tick can be wrong, referring to the fact that a tick will suck a liquid that will kill it as long as all the other signals have been triggered – the correct temperature of the liquid, for example. For Whitehead, however, what is important is the sense in which it is not the immediately given signal (e.g. the odour of butyric acid) that prompts the action, but rather it is the symbolic referent (which is distinct from what is given in presentational immediacy) that is the prompt to action. But how do we differentiate the two? Is not the tick responding to the butyric acid as a signal that refers to the presence of a warm-blooded mammal, much as Aesop's dog drops the meat in its mouth in response to the meat it infers as the cause of what it sees reflected in the water? For Whitehead what differentiates the two cases, and then more elaborately so with linguistically expressed thoughts and the advent of indirect communication, is that the 'reflex action' of instinct 'is wholly dependent on sense-presentation, unaccompanied by any analysis of causal efficacy via symbolic reference' (S 81), and it is the '*conscious* analysis of perception' (emphasis added) that differentiates the tick from the dog, as well as from the human who expresses thoughts in words.

At this point have we not opened another Pandora's box and raised a whole host of new questions regarding the nature of consciousness? If it is the 'conscious analysis of perception' that differentiates the instinctual, reflex action from the symbolically conditioned action, then what is the nature of consciousness? To answer as briefly as possible, and then slowly to return to the questions regarding the differences between ticks, dogs and humans, let me begin by offering a Sartrean definition of consciousness as consisting in the deterritorialisation of established social norms and habits. As Sartre argued that consciousness is consciousness of something to the extent that it is not that which it is conscious of, similarly it can be argued that consciousness deterritorialises that which is, and in doing so also entails the process of reterritorialisation that leads to the emergence of something other. It is this something other that is the underlying referent of symbolic inferences and the sufficient reason for consciousness of something at all, and for the new habits and social forms that emerge as the deterritorialised processes of consciousness become reterritorialised. What is integral to the nature of consciousness, therefore, is precisely the dual process of deterritorialisation and reterritorialisation.

Symbolic reference, for example, entails the deterritorialisation of presentational immediacy, or 'sense-presentation' as Whitehead puts it, which in turn becomes reterritorialised as the object or other that is inferred to be the basis or causal efficacy of that which is given in 'sense-presentation'. Moving to organisms more generally, consciousness is the deterritorialising subset of habits, retentions and expectations that constitute the individuality of an organism. This subset is not to be identified with any determinate set of habits, retentions and expectations, for then the subset would be a territorialised part of the organism; to the contrary, consciousness is the subset that is irreducible to any determinate set or subset of habits, and yet it is nothing over and above them. As Deleuze lays out this relationship in *Difference and Repetition*, this deterritorialising subset is the plane of consistency that is irreducible to the determinate elements and spatio-temporal powers and habits that actually develop and come to be identified with the individuated entity. In *What Is Philosophy?*, for example, Deleuze and Guattari argue that philosophical concepts are always composed of components, but what is critical to the concept is not the components themselves but rather, as they put it, the 'area *ab* that belongs to both [components] *a* and *b*, where *a* and *b* "become" indiscernible. These zones, thresholds, or becomings, this inseparability, define the internal consistency of the concept'.[28] It is this internal consistency of the concept that is the life or becoming of the concept. This consistency is precisely the deterritorialised subset of the components that tends towards the deterritorialising pole of chaos – what we will call the panvitalist pole, in that this is the life of the components, the deterritorialising life of thought. Yet this subset is inseparable from the identity and individuality of the determinate components, components that are themselves the result of the unifying, reterritorialising processes of contemplation and thought – what we will call the panpsychic pole. Consciousness, therefore, or thought properly so-called, is, to use Whitehead's terminology, an event or actual occasion, in that it involves the dual tendencies of deterritorialisation and reterritorialisation, tendencies that are inseparable from each and every actual thing. Every actual thing, in short, tends towards both its panvitalist and panpsychic poles.

To begin to clarify the sense in which thought and consciousness themselves are nothing less than the dual tendencies of deterritorialisation and reterritorialisation, it will help to turn to Deleuze and Guattari's admission that their project is vitalist. Deleuze and Guattari do this in the conclusion to *What Is Philosophy?*, which is significant given the retrospective nature of this book. Deleuze and Guattari are careful to note, however, that vitalism 'has always had two possible interpretations'. On the one hand there is vitalism as a life or 'Idea that acts, but is not'.[29] This is the sense that can be given to Whitehead's *vitalism*, for, as he argues, each and every actual thing is a society with an instinct that *acts* to synthesise and draw together the relevant occasions into a synthesis. As Whitehead argues, the individuality of each and every actual thing 'consists in its synthesis of other things so far as they are relevant to it' (S 26). This synthesis that gives rise to the individuality of actual things is made possible, however, by the vitalism that is understood as 'a force that is but does not act – that is therefore a pure internal Awareness'.[30] It is this vitalism that Deleuze and Guattari take 'to be imperative',[31] for it is precisely what the active syntheses presuppose. Deleuze and Guattari's argument for this claim closely follows and extends Hume's philosophy. In particular, they argue for a contraction of habits, or passive synthesis, that must 'be contracted in a contemplating "imagination" while remaining distinct in relation to actions and to knowledge'.[32] In other words, what is stressed here is the panvitalist tendency where what is crucial is not the active synthesis and becomings of determinate and actual things (the panpsychic, reterritorialising tendency), but rather the consistency that allows for these becomings in the first place. As Deleuze and Guattari put it, what we must find 'beneath the noise of actions, [are] those internal creative sensations or those silent contemplations that bear witness to a brain', and it is precisely a Humean vitalism that is critical to Deleuze and Guattari's project in *What Is Philosophy?* and to Deleuze's project more generally.[33]

In referencing the 'silent contemplations that bear witness to a brain', Deleuze and Guattari have in mind the Humean argument, extended by Deleuze in his own work, whereby elements are brought together into a state of consistency without there being an active, determinant role at work in the synthesising process. This process is what Deleuze called 'passive synthesis' in *Difference and Repetition*, and it is Hume who sets the stage for Deleuze's arguments. Moreover, much of the work that Deleuze and Guattari did together can be seen as an extension of these arguments, as is largely acknowledged in the conclusion to *What Is Philosophy?* when Deleuze and Guattari invoke Plotinus's claim that 'all is contemplation' in order to justify their vitalism, or what they call 'an inorganic life of things'.[34] The Humean contemplations are precisely the passive syntheses that are external to the elements synthesised and which make active syntheses possible as these passive syntheses are picked up by an active synthesis. For example, Deleuze offers the example of thirst to illustrate the interplay between passive and active synthesis. The key move in understanding this interplay is the manner in which Deleuze extends Humean contemplation and habit formation to all entities, including, but not limited to, all organisms. Deleuze is quite forthright on this point: 'We are made of contracted water, earth, light and air . . . Every organism, in its receptive and perceptual elements, but also in its viscera, is a sum of contractions, of retentions and expectations'.[35] Deleuze then adds that 'each contraction, each passive synthesis, constitutes a sign which is interpreted or deployed in active synthesis'.[36] The interplay between passive and active synthesis, therefore, is that between the constitution of signs and their deployment in an active synthesis. In the case of thirst, for instance, this state involves the contemplation, C, or passive synthesis and contraction of elements that constitute the sign that then allows an animal to seek the presence of water by way of an activity, A. To restate this in the context of Deleuze and Guattari's claim that we need 'to discover, beneath the noise of actions, those internal creative sensations or those silent contemplations', we can say that beneath the action of seeking water, A, there are the passive syntheses, contractions and contemplations of elements, C, that constitute the sign that is then taken up by A. In the conclusion to *What Is Philosophy?*, Deleuze and Guattari refer to this contemplation of elements, C, that allows for the possibility of active, determinate relations, A, as the brain. The 'silent contemplations . . . bear witness to a brain', Deleuze and Guattari claim.

In Deleuze and Guattari's discussion of the brain, the brain is not to be confused with the bodily organ; rather, the brain is the process of contraction and contemplation, or C, that makes it possible for a determinate subject to have a determinate thought, or A. It is for this reason that Deleuze and Guattari claim that 'It is the brain that thinks and not man – the latter being only a cerebral crystallization'.[37] Deleuze and Guattari add that they 'will speak of the brain as Cézanne spoke of the landscape: man absent from, but completely within the brain'.[38] In other words, the active thinking subject is not to be confused with the contemplations and passive syntheses that made this subject possible. The subject is completely within the brain, therefore, for the subject is nothing over and above the process of contemplation and contraction that constitutes the signs that are then taken up in processes of active synthesis. The same is the case for plants, Deleuze and Guattari argue, as the 'plant contemplates by contracting the elements from which it originates – light, carbon, and the salts – and it fills itself with colors and odors that in each case qualify its variety, its composition: it is sensation in itself'.[39] The plant thus draws together, in a passive synthesis, a subset of heterogeneous elements such as water, nitrogen, carbon, chlorides and sulphates that are then drawn into a plane of consistency that allows for the emergence of signs that are then taken up by the active syntheses of the plant, such as its being phototropic. These active syntheses can themselves be taken up as the signs – such as the 'colors and odors' of flowers – that lead to the active syntheses and behaviours of wasps, bees and birds, and so on. What is key to this process for Deleuze and Guattari is that a plane of consistency be drawn on a plane of immanence, the latter being, Deleuze and Guattari claim, 'like a section [that is, subset] of chaos and acts . . . as a sieve'.[40] Humean contemplations and passive syntheses thus draw upon a plane of immanence the elements that are a subset of chaos, and the plane of consistency is then taken up and leads to a new active synthesis. The brain, in other words, is the plane of immanence that has already filtered chaos, served as a sieve, and it is the brain that is presupposed by the processes of passive synthesis which allows for the possibility of active processes of individuation and for the determinate relations between individuals. It is this view of the brain that leads Deleuze and Guattari to the final formulation of their vitalism:

> Not every organism has a brain, and not all life is organic, but everywhere there are forces that constitute microbrains, or an inorganic life of things.[41]

In Deleuze and Guattari's version of Plotinus's claim that 'all is contemplation', therefore, they argue that everywhere there are microbrains or 'an inorganic life of things', and this inorganic life is not to be confused with the determinate lives of animal species, or even with organic life itself; rather, this life is precisely the passive syntheses and contraction of elements – that is, the contemplations – that make possible the emergence and individuation of determinate entities, whether organic or not, living or not. The microbrains, in short, are the very life of things, the life that allows for the very determinate existence of things at all.

We can now return to our earlier question. In light of what we have argued, a tick is simply the sum of its reterritorialised habits and it is not capable of breaking with habits in light of what happens on any given occasion. The behaviour of a tick, therefore, consists solely of taking up the signs (butyric acid etc.) that were constituted as a result of passive synthesis or contemplation, and it does so automatically. There is no room for interpretation of these signs, no room for hesitation or error; to the contrary, the behaviour of the tick is fully reterritorialised upon the sign itself, and thus given the appropriate sign the tick will suck a poisonous liquid. This exemplifies the panpsychic tendency, the tendency to recollection, unity and reterritorialisation. Moving on to a dog, it is not fully reterritorialised upon its habits and is capable of deterritorialising established habits and modifying its behaviour as it responds to and learns from changes in its environment. There are limits to the thoughts dogs can have, as we have seen, in that their thoughts and behaviours are linked to actual stimuli. There is thus limited room for deterritorialising thoughts and behaviours in the case of dogs. The ability to employ tools and instrumental reasoning, however, allows for a further deterritorialisation of established behaviours and habits, which then become reterritorialised upon tools and technologies. As André Leroi-Gourhan has argued, when early humans began to move from their habitat in the trees, their hands became deterritorialised from their established habits and behaviours, only to become reterritorialised upon the use of tools; similarly, the mouth was deterritorialised from its role in assisting in keeping a group together as they move through trees and became reterritorialised upon speech.[42] And finally, the ability to express thoughts through syntactically structured language deterritorialises thoughts from their relationship to practical contexts of usage and reterritorialises them upon their relationship to other thoughts, and with this we have the emergence of indirect discourse. It is this de/reterritorialisation process that constitutes the nature of thought or consciousness, in the Humean/Plotinian sense, and it is the necessity of maintaining both tendencies that is critical to Whitehead's philosophy of organism.

This latter point becomes clear in Whitehead's critique of the tendency for symbols to become automatic reflex actions. When this occurs, Whitehead claims that what has happened is that an 'organism has acquired the habit of action in response to immediate sense-perception, and has discarded the symbolic enhancement of causal efficacy' (S 81). In other words, rather than infer to the other to which symbolic reference directs us, the symbol itself functions much as an instinct for a tick – to wit, our response becomes an automatic, reflex behaviour. Since symbolic reference may well be wrong, a symbolic action that has become reterritorialised as a reflex action may well be an action determined by a false ideology, a false assumption, such as African-Americans in hoodies are a danger or a threat, the government is always wasteful, and markets are always free, efficient and good, and so on. What is needed, therefore, is precisely the work of thought in order to deterritorialise such habits and reflex actions. We must not go too far in this deterritorialisation process, however, or we must maintain the plane of immanence, for otherwise we risk collapsing into the chaos that undermines the very possibility of maintaining the social multiplicity that constitutes an actual thing. A political system, Whitehead warns, must not change all the habits and established patterns, for a change or revolution that is too dramatic will likely 'be destructive of the social system' (S 71). Life in Virginia, for instance, to refer to Whitehead's example, was not destroyed, for 'the prejudices on which Virginian society depended were unbroken. The ordinary signs still beckoned people to their ordinary actions, and suggested the ordinary common-sense justification' (S 77). For Whitehead an analysis and deterritorialisation of established habits and customs are thus essential for thought, and yet this process must maintain a foothold within established habits and customs. To state the point using Whitehead's terms, thought consists of nothing less than events or actual occasions, and this entails that thoughts presuppose, as limits to be avoided, the panpsychic pole of absolute unity and reterritorialisation and the panvitalist pole of deterritorialising chaos. As Whitehead describes his philosophy, it is a philosophy of organism, and an organism, moreover, avoids fully actualising both the panvitalist and panpsychic poles. Whitehead's *Symbolism: Its Meaning and Effect* provides us with a succinct yet thorough account of this philosophy.

Notes

1. Aristotle, *The Politics*, I.2.
2. Descartes, letter to Henry More of 5 February 1649. See Adam and Tannery, *Oeuvres de Descartes*, vol. IV, 574.
3. Adam and Tannery, *Oeuvres de Descartes*, vol. V, 276–7.

4 In his *Enneads*, Plotinus argued that 'all is contemplation'. Deleuze and Deleuze and Guattari will adopt this theme, as we will see in section III of this chapter.
5 It should be noted that in the commentary on Whitehead the terms vitalism, panpsychism and panexperientialism have been used from nearly the beginning. Charles Hartshorne, for instance, defends a panpsychic reading of Whitehead (Hartshorne, 'On some criticisms of Whitehead's philosophy'). These terms have continued to be used by many other scholars who write under the influence of Whitehead. The use of these terms continues to be a matter of debate and discussion among Whitehead's commentators, as is evidenced within the collection of essays in this volume.
6 See Hauser, *Wild Minds*; Hauser et al., 'The faculty of language'; Griffin, *Animal Minds*.
7 Camp, 'Putting thoughts to work', 276.
8 Camp, 'Putting thoughts to work', 276.
9 Camp, 'Putting thoughts to work', 280.
10 Camp, 'Putting thoughts to work', 280.
11 Camp, 'Putting thoughts to work', 280.
12 Camp, 'Putting thoughts to work', 278, citing Evans, *The Varieties of Reference*, 104.
13 For Sellars' critique of the 'Myth of the Given', see his *Empiricism and the Philosophy of Mind*. See also McDowell, *Mind and World*, and Brandom, *Articulating Reasons*.
14 Whitehead is presupposing here the important arguments from his *Process and Reality*.
15 von Uexküll, *A Foray into the Worlds of Animals and Humans*, 50.
16 For example, a tick will suck up the blood if it is not the right temperature but it will suck up a liquid that kills it if it is at the right temperature. See von Uexküll, *A Foray into the Worlds of Animals and Humans*, 44–53.
17 Camp, 'Putting thoughts to work', 295.
18 Camp, 'Putting thoughts to work', 296.
19 Camp, 'Putting thoughts to work', 296.
20 See Gould and Gould, *The Honey Bee*.
21 Camp, 'Putting thoughts to work', 299.
22 Camp, 'Putting thoughts to work', 299.
23 Camp, 'Putting thoughts to work', 301.
24 Camp, 'Putting thoughts to work', 301.
25 Camp, 'Putting thoughts to work', 304.
26 See Benveniste, *Problems in General Linguistics*.
27 Camp, 'Putting thoughts to work', 304.
28 Deleuze and Guattari, *What Is Philosophy?*, 19–20.
29 Deleuze and Guattari, *What Is Philosophy?*, 213.

30 Deleuze and Guattari, *What Is Philosophy?*, 213.
31 Deleuze and Guattari, *What Is Philosophy?*, 213.
32 Deleuze and Guattari, *What Is Philosophy?*, 213.
33 See my *Deleuze's Hume*, as well as my *Deleuze and Guattari's* What Is Philosophy?
34 Deleuze and Guattari, *What Is Philosophy?*, 213.
35 Deleuze, *Difference and Repetition*, 73.
36 Deleuze, *Difference and Repetition*, 73.
37 Deleuze and Guattari, *What Is Philosophy?*, 12.
38 Deleuze and Guattari, *What Is Philosophy?*, 212.
39 Deleuze and Guattari, *What Is Philosophy?*, 212.
40 Deleuze and Guattari, *What Is Philosophy?*, 42. There is a close parallel with Deleuze's understanding of the relationship between passive and active synthesis and Jessica Wilson's use (see Wilson, 'Non-reductive realization and the powers-based subset strategy') of what she calls the powers-based subset strategy to lay out a theory of mental phenomena that is non-reductive to physical phenomena and yet nothing over and above these physical phenomena. As she argues, mental phenomena can be understood non-reductively if they are a proper subset of physical powers. To restate Deleuze's example of thirst in Wilson's terms, the mental phenomenon of thirst is a determinable subset of determinate powers, and what is crucial is the relevance of the subset as a sign that is then taken up by the action of seeking water. The determinable subset is thus multiply realisable, in that a different subset of heterogeneous elements can be taken up as a sign for the same active processes (e.g. searching for water). For Deleuze, similarly, the signs constituted by contemplation and passive synthesis are also multiply realisable upon the plane of immanence. The plane of immanence or sieve, in other words, can draw from many different elements; what is important is the relevance of the sign as it is taken up by an active synthesis.
41 Deleuze and Guattari, *What Is Philosophy?*, 213.
42 Leroi-Gourhan, *Gesture and Speech*.

Bibliography

Adam, Charles, and Paul Tannery, *Oeuvres de Descartes*, vols I–XII, revised edition (Paris: J. Vrin/CNRS, 1964–76).

Aristotle, *The Politics and The Constitution of Athens*, trans. Stephen Everson (Cambridge: Cambridge University Press, 1996).

Bell, Jeffrey A., *Deleuze and Guattari's* What Is Philosophy? *A Critical Introduction and Guide* (Edinburgh: Edinburgh University Press, 2016).

Bell, Jeffrey A., *Deleuze's Hume: Philosophy, Culture and the Scottish Enlightenment* (Edinburgh: Edinburgh University Press, 2009).

Benveniste, Émile, *Problems in General Linguistics*, trans. Mary Elizabeth Meek (Coral Gables: University of Florida Press, 1966–74).

Brandom, Robert B., *Articulating Reasons: An Introduction to Inferentialism* (Cambridge, MA: Harvard University Press, 2000).

Camp, Elisabeth, 'Putting thoughts to work: concepts, systematicity, and stimulus-independence', *Philosophy and Phenomenological Research*, 78:2 (2009), 275–311.

Deleuze, Gilles, *Difference and Repetition*, trans. Paul Patton (New York: Columbia University Press, 1994).

Deleuze, Gilles, and Félix Guattari, *What Is Philosophy?* (New York: Columbia University Press, 1994).

Evans, Gareth, *The Varieties of Reference* (Oxford: Oxford University Press, 1982).

Gould, James, and Carol Grant Gould, *The Honey Bee* (New York: Science American Library, 1988).

Griffin, Donald R., *Animal Minds* (Chicago: University of Chicago Press, 1992).

Hartshorne, Charles, 'On some criticisms of Whitehead's philosophy', *Philosophical Review*, 44:4 (1935), 323–44.

Hauser, Marc D., *Wild Minds: What Animals Really Think* (New York: Henry Holt and Company, 2001).

Hauser, Marc D., Noam Chomsky and Tecumseh Fitch, 'The faculty of language: what is it, who has it, and how did it evolve?', *Science*, 298 (2002), 1569–79.

Leroi-Gourhan, André, *Gesture and Speech* (Cambridge, MA: MIT Press, 1993).

McDowell, John, *Mind and World* (Cambridge, MA: Harvard University Press, 1994).

Sellars, Wilfrid, *Empiricism and the Philosophy of Mind* (Cambridge, MA: Harvard University Press, [1956] 1997).

Uexküll, Jakob von, *A Foray into the Worlds of Animals and Humans: With a Theory of Meaning* (Minneapolis: Posthumanities Press, 2010).

Whitehead, Alfred North, *Symbolism: Its Meaning and Effect* (New York: Capricorn Books, [1927] 1959).

Wilson, Jessica M., 'Non-reductive realization and the powers-based subset strategy', *The Monist (Issue on Powers)*, 94:1 (2011), 121–54.

9

From Manipulation to Co-creation: Whitehead on the Ethics of Symbol-Making

LUKE B. HIGGINS

Twenty-first-century humanity seems to have both far too much and far too little invested in symbolism. On one hand, the culture of late capitalism seems obsessively intent on insinuating increasingly manipulative and hard-to-ignore symbols – on increasingly diverse platforms – into the inner sanctum of our lives. In this sense, the human world is a veritable tangled thicket of symbols gone wild, one that literally calls out for Alfred North Whitehead's injunction to engage in a 'continuous process of pruning' so as to keep from being 'overwhelmed by our symbolic accessories' (S 61). It seems that everywhere we turn someone is trying to manipulate us with symbols.

On the other hand, it seems equally true that we are starved for the *kinds* of symbols that can provide real sustenance and connective power for the hard work we face of rescuing our vulnerable planet from imminent disaster. One might even ask whether our growing cynicism in the face of the manipulative symbols of the market has damaged our capacity to invest ourselves in more worthwhile symbols. Given that most of us have honed our critical thinking skills in the crucible of predatory market capitalism, is it any wonder that we feel cynical in the face of so many attempts to symbolise our higher ideals, whether they appear in the form of religious affirmations of universal love or ethico-political visions of global ecological unity? It does not help that justice-pursuing organisations of various kinds (including religious ones) feel compelled to resort to many of the same marketing techniques that are used to sell commercial products. By this logic, we might even be able to comprehend that bizarre conservative delusion that contemporary environmentalist calls to care for the health of our living biosphere originate in a liberal conspiratorial hoax to bolster the tyrannical power of 'big government'.

In this cultural milieu it would seem vastly preferable to position oneself as a manipula*tor* of symbols rather than submit to being manipulated *by* them. Within the terms of this perhaps more cynical but also more practical logic, the meanings of our symbols would ultimately be reducible to some underlying set of knowable realities. For example, evolutionary psychology informs us that underneath the symbolic affirmations of religion, we find little more than an obsolete instrument of natural selection aimed at binding social groups together. (As we shall see, there *is* a kind of half-truth expressed here, which adds to its apparent credibility.) This line of thinking leads us to assume that the only helpful symbols are those that can be easily cashed out into their non-symbolic, factual content. In Alfred North Whitehead's words, 'Hard-headed men want facts and not symbols', and hard science is appointed with the all but sacred task in our society of systematically weeding out the latter from the former (a process Bruno Latour calls 'purification' in his work *We Have Never Been Modern*).[1] Once nature has been reduced to these all-too 'handy' (as Whitehead calls them) tools, there is little in our world that cannot be exploited as grist for those twin mills of our contemporary civilisation – technoscience and capitalism.

In this chapter I wish to invoke the possibility of a new approach to symbolism that moves beyond two fundamentally sceptical assumptions regarding meaning-making. The first is that the only appropriate use of symbols is to exploit their ability to reduce our world to a set of abstractly manipulable symbolic units, that is, the 'laws of nature', or – as the case may be – the laws of economics (which can be even more ruthless in their exploitive logic of value extraction). In other words, the only 'good' symbols are really just signs that have an unproblematic correlation with their signified objects. The second assumption is that any meaning that cannot be 'cashed out' of its symbolic form is fundamentally illusory and thus manipulative – whether this symbolic meaning is 'projected' onto a traditionally conceived divinity, or cynically attributed to the machinations of politico-economic power. I would submit that Whitehead's analysis of symbolism and its role in perception and creative thought/feeling might constitute the basis for such a re-envisioning of the role of symbolism in our current 'natureculture' (to use a neologism coined by Donna Haraway).

We will find that while the operation of what Whitehead calls 'symbolic reference' creatively forges a vast web of associations that underlie both perceptive and emotive experience, it plays just as central a role in the ossification and even reification of these connections into habituated 'reflexes'. The latter goes a long way towards explaining not just the utility of symbols in the machinations of social-political power, but also the common assumption that certain symbols exist 'ready-made' – either in a transcendently divine sphere, or as the scientific 'master-code' to which we can reduce the operations of nature. In the case of the latter, of course, it is precisely the symbolic status of scientific facts that is denied, leading to a bifurcated worldview where the symbols that *do* exist (religious or otherwise) are seen as little more than mental/cultural projections that aesthetically – and sometimes ethically – adorn an otherwise static, mechanistic reality.

I submit that there is indeed a third way of engaging symbolism, which goes beyond either using symbols to manipulate objects (or people) or submitting oneself to the manipulations of some transcendently conceived symbol-complex. For both these alternatives, in my view, take their bearing from a more or less fearful response to that basic ontological reality that Whitehead calls 'evil': namely, the mutually obstructive character of the universe, in which the greater the novelty and significance achieved, the greater its tendency to 'fade into night' (PR 341). A third path, then, might take its cue from some of Whitehead's more enigmatic reflections at the end of both *Process and Reality* and *Adventures of Ideas*, in which he signals towards a capacity to draw from a deeper, divine stratum of the world's creative process, where novelty does not mean loss. Whitehead's almost mystical intimations here may point us towards a unique mode of symbol-making (or symbol-revision, as the case may be) that emerges in the between-space of our individual self-creativity and our immersion in a world that exceeds us in every way. It would aim at both receiving and transmitting the power of something like what Gilles Deleuze calls 'a life' – that unique quality of the living that is universal in its singularity.

To summarise the thesis of this chapter: symbols are not merely fabulations of our imagination. Neither are they pre-existent, either within a transcendent divine subject or as ready-made tools with which to master nature's objects. Rather, symbols are *produced* in and through the indeterminate and dynamically ecological relations of creative becoming. Moreover, this symbolic production comes to *constitute the real connective fabric of the world*, the sustenance of which – at this point – is a matter of life and death for our entire earth collective. Thus, the ultimate question of symbolism shifts from a metaphysical or epistemological register to a more fundamentally *ethical* one: how should we take up the challenge of living as symbol-makers? (I use the term 'ethical' primarily in the Spinozist sense of enquiring into the powers of which we are capable, rather than the laws to which we should submit.) And from what reserves do we draw the courage, resolve and inspiration to give ourselves over to this creative work?[2]

The Role of Symbol-Making in Perception and Pragmatic Action

For Whitehead, the domain marked out by the relationship between a symbol and its meaning is an extremely broad one: symbolic functioning takes place any time one component of experience 'elicits consciousness, beliefs, emotions and usages, respecting other components of experience' (S 8). However, by far the most pervasive use of symbolic reference is the one that correlates sense-presentation with physical bodies, resulting in the animal experience of perception. As many other authors in this volume have pointed out, what distinguishes Whitehead's analysis of perception from so many others is his insistence that it is dually constituted – that is, there are two fundamental components that go into any perceptive experience: 'causal efficacy' proceeds *from* the world of our settled past *to* the self-creative emergence of our subjectivity, and defines the sense in which all actually existing entities are (partially) constituted by the vectorial waves of propulsion that precede them. As we shall see, these waves are not blindly mechanistic in their flow, but are inherently productive of what Whitehead calls 'feeling'.

While causal efficacy is the major constitutive component of both inorganic and 'lower' organic societies, a significant threshold is crossed in the evolutionary transition between these two. In the case of the latter, a more complex set of 'interested' strategies for receiving and relating to a larger world comes into effect. Organic societies thus respond to their world with a far greater sense of purpose than inorganic ones, but compared with 'higher' animals, there is little hesitation in deciding how to respond to the world. Whitehead sums it up well: 'A flower turns toward the light with much greater certainty than does a human being, and a stone conforms to the conditions set by its external environment with much greater certainty than does a flower' (S 42).

Animals with more complex, centralised nervous systems originate a mode of relational becoming which is more individually purposeful, yet also more prone to both hesitation and – for the first time – error.

This is all made possible by the enhancement of another mode of interaction that both overlays causal efficacy and proceeds in the opposite direction – that is, it moves *from* the living entity in question *to* particular aspects of (what it imagines to be) its contemporary, external world. 'Presentational immediacy' employs the supplemental phases of concrescence to 'project' (although Whitehead insists on using this word only in a provisional sense) particular sense qualities on a world accessed through the specialised operations of the sense organs. As opposed to the vague, emotionally laden necessities of causal efficacy, presentational immediacy is vivid, distinct and – in itself anyway – empty of any inherent significance. While causal efficacy expresses the bare 'givenness' of a world that must be inherited, presentational immediacy gives rise to sense impressions whose correlation with an actual, external reality must necessarily remain indeterminate. The maze of epistemological problems with which Enlightenment philosophers such as Locke, Hume and Kant wrestled is, for Whitehead, very neatly explained by the recognition that they considered *only* the evidence of presentational immediacy, nearly ignoring causal efficacy altogether.

The first and most important insight to be drawn from all this is that the human being's most primary and significant application of symbolic reference – namely, perception – can be reduced neither to a pre-existent order, passively received, nor to an imaginative fabrication, actively imposed. Symbolic reference is *necessarily* the product of a synthetic correlation between the two, and there is no fool-proof, preordained formula that can guarantee the 'accuracy' of some particular mode of correlation. In Whitehead's words:

> the symbolic reference leads to a transference of emotion, purpose and belief, which cannot be justified by an intellectual comparison of the direct information from the two schemas and their elements of intersection. The justification, such as it is, must be sought in a pragmatic appeal to the future. (S 30–1)

Symbol-making – which underlies not just perception, but also language, along with the wildly diverse meaning systems on which humanity's social-cultural life is based – has its basis in a *creative synthesis* between an interested, purposeful subjectivity and a world within which variously complex orders inhere. We are neither the sole authors nor the sole inheritors of our experience. Even more significantly, there is no way of neatly sorting out which aspects of our reality are objectively given and which are subjectively 'projected'.

In a radical departure from modernity's demand that we dichotomise or 'bifurcate' (as Whitehead puts it) between the facts of nature and the values of the human mind and culture, Whitehead insists that there *are no* meanings that do not always already exist as a synthesis of both. As we shall discuss further below, it is precisely because of this that a certain risk of error – not mention outright failure – is impossible to expunge from our basic constitution as human beings.

Nevertheless, it is most definitely true that symbolic reference opens a path for living organisms to take a far more active and creatively spontaneous role in the 'invention' of their reality. Whereas in earlier phases of development organisms functioned primarily as passive, instinctive vessels for a set of purposes mostly preordained by the species history, now for the first time individual living beings are able to grasp their world as something that can be *acted upon* in a number of different possible ways. By drawing correlations between either *common* sensory experiences (given by presentational immediacy) over different times/places (given by causal efficacy) or *diverse* sensory experiences converging in *one* time/place (PR 170–1), we can interact not just with the vague waves of input from the immediate past, but now with the more vivid patterns of a world whose complexities present opportunities for our future. This brings us to our second major insight from Whitehead's analysis: the basic patterns of human experience are integrally shaped by the specific strategies evolved by the human organism to selectively interact with those features of the world that have some kind of importance or pragmatic value. In the following quote, Whitehead observes that, although we *could* absorb ourselves in the present immediacy of the sensory impressions made by a coloured traffic light (observing that it is red and so on), what is really pressing for us is the way it symbolises the set of causal forces determining our future:

> The symbols do not create their meaning: the meaning, in the form of actual effective beings reacting upon us, exists for us in its own right. But the symbols discover this meaning for us. They discover it because in the long course of adaptation of living organisms to their environment, nature taught their use. It developed us so that our projected sensations indicate in general those regions which are the seat of important organisms. (S 57)

As Whitehead develops his theory of perception further in *Process and Reality*, he discusses how presentational immediacy displays not the *actual* world, which for us can only be the actuality of our immediate past, but rather an abstract image of the way our contemporary world could potentially be divided.

> Presentational immediacy illustrates the contemporary world in respect to its potentiality for extensive subdivision into atomic actualities and in respect to the scheme of perspective relationships which thereby eventuate. But it gives no information as to the actual atomization of the contemporary 'real potentiality'. (PR 123)

A more faithful interpretation of this quote's meaning would require a grasp of Whitehead's complex theory of strains and loci – ideas that reach beyond the purview of this chapter. I invoke it here to emphasise the sense in which presentational immediacy overlays the actualities imposed on us by causal efficacy with a layer of perceived potentiality. Through these complex operations (which include what Whitehead calls the 'category of transmutation'), a higher-grade organism is able to 'eliminate, by negative prehensions, the irrelevant accidents in its environment', thus 'eliciting massive attention to every variety of systemic order' (PR 317). This is to say that what an actively intelligent mind takes in of the external world is strictly filtered according to the way the world's structured orders submit themselves – or fail to submit themselves – to a range of potential actions.

It is important then to realise that, despite Whitehead's philosophical realism, perception for him is not designed to provide unbiased information about our world. Rather, what we have access to in perception is a symbolic network aimed at highlighting features of the world that hold potential importance or value for our species. In short, perception yields not the world in itself, but a *world of interest*. For instance, objects in motion in our environment tend to stand out because they signal events that have a greater potentiality for either imminently affecting or being affected by our own mobile bodies. Perception tends to edit out those diversities of detail that compromise our attunement to certain abstract commonalities among objects. This 'automatic' operation is what allows us to use past experiences to interpret present ones: if a flat surface of orange rectangles signified a brick wall in the past, a similar presentation of sense qualities is likely to signify a brick wall in the present.

Once some particular symbolic complex has been established to have an ongoing pragmatic relevance, the need to ponder the relationship between symbol and meaning lessens. The symbol is able to 'conjure up' the action or practice tied to its meaning in a habituated, reflexive manner. These kinds of shortcuts in our experience of the world have an irreplaceable utility for us – if we were unable to move through our world to a great extent 'on automatic', we would not have either the time or the energy to devote ourselves to the more challenging tasks and problems of life. Nevertheless, the fact that we do deploy so many reflexive shortcuts in our experience of the world can lead to the mistaken assumption that what 'really exists' out there *is* the symbolically coded action (or potential for action) that it subjectively evokes in us. For these reasons, it is quite natural for human beings to experience a universe at which we are the centre – either as individuals or as a species.

Although it is beyond the purview of this chapter to trace out the following connections as fully as I might, I would suggest that modern technoscience is, in many ways, an extension on a larger scale of some of these same tendencies. Scientific facts take to the extreme the idea that our world can be reduced to an underlying code, endlessly manipulable. Of course, the pursuit of these so-called 'facts' is framed as precisely an overcoming rather than a construction of symbols: 'A clear theoretic intellect, with its generous enthusiasm for the exact truth at all costs and hazards, pushes aside symbols as being mere make-believes, veiling and distorting that inner sanctuary of simple truth that reason claims as its own' (S 60). Nevertheless, the very utility of scientific symbols – as Whitehead puts it, their 'handyness' – lies in the way they congeal a unit of associative information such that it can be pragmatically mapped onto other, similar regions of experience.

This is related to what Whitehead observes as modern science's deep and problematic investment in the 'fallacy of misplaced concreteness' – namely, the tendency to mistake an abstract concept for the actual, concrete reality from which it was derived. In *Process and Reality*, Whitehead explains how this basic process of dissimulation takes place in scientific research and theorising. On the one hand, scientific observations are made exclusively in the mode of presentational immediacy – that is, there is an attempt to 'purge' observations of the larger width of experience (given by causal efficacy) that give them their specificity and individual character. This process is intended to systematically eliminate any remaining element of subjective interpretation. On the other hand, in Whitehead's words, 'all scientific theory is stated in terms referring exclusively to the scheme of relatedness, which, so far as it is observed, involves the percepta in the pure mode of causal efficacy' (PR 169). This sleight of hand whereby the content of one mode of perception is traded out for the other is part of how science is able to attribute a changeless determinism to the facts of nature that the unpredictable flux of the real world does not actually possess.

My point here is that the fallacies of modern science are partly derived from the illusions to which human perception naturally gives rise. Ironically, however, as we become increasingly invested as a culture in a world of facts bestowed upon us by scientific experts, we also tend to become increasingly withdrawn from that richly textured natural world to which our senses give access – a strict cultural application of empiricism thus comes to undermine our *actual* empirical faculties. That is, by habitually assuming that the meaning of things lies primarily behind the surface, we give less attention to the living, sensory world within which our earlier ancestors were unavoidably immersed, a point that David Abram makes quite provocatively in his book *The Spell of the Sensuous*.[3] Indeed, technoscientific processes have been so wildly successful at mediating our relationship with the non-human world that the average human being comes to have few pragmatic reasons to directly engage with it. Tempted by the convenience of relying simply on the master-code of the experts, we forfeit our own unique destiny as symbol-makers. We make reflexive assumptions about the meaning to which a word-symbol like 'forest' points, instead of taking up the challenge of the poet described by Whitehead, who encounters the forest directly to see what new symbolic meanings it might give rise to.

This is not to underplay the countless advantages provided by these handy symbolic tools. They are what enable us to actively shape our environment in ways that support the diverse and often wondrous activities of human life and culture. However, as we become increasingly habituated to a static set of symbol–action associations – in other words, as they start to function reflexively for us – we become prone to losing track of the ways our experience of reality is always symbolically synthesised from the ground up. This also means that we lose touch with our own innate potential to re-synthesise or recreate that symbolic reality.

A number of difficulties, delusions and even existential crises for humanity follow in the wake of this forgetting. One of these is the tendency to reify the separation between our own individual selves and the diverse becomings of the world around us. That is, while the capacity to reduce our world to its pragmatic value opens up possibilities for our own becoming, it also alienates us from a sense of being a part of something bigger. As beings co-constituted by presentational immediacy, our experience situates us primarily as a subject in a world of objects. On the other hand, living organisms whose experience is dominated primarily by causal efficacy would presumably experience themselves as a subject – though a much more vague and diffuse one to be sure – swept up in the same currents of becoming as every other subject. We tend to assume that, as beings with highly developed sensory capacities and consciousness, we experience *more* of our world. But that is true only in a certain sense. We experience more of the abstract set of relations that potentially connect and dissect our world, but this comes at the cost of filtering out a more direct sense of 'kinship' with our world. The selectivity that our senses impose on our relations with the world also alienates us from an (admittedly vague) awareness of participating in a flow of becoming that exceeds our individuality.

Human civilisation today can more or less be defined as the institutionalisation of this strategic reduction whereby subjects pragmatically utilise symbolically reduced objects. The relationship we pursue with our environment is increasingly reduced to a mere processing of raw material for the engines of capitalism. Moreover, the logic of capitalism applies these same strategic principles to other human beings, strategically reducing them to labourers, consumers or – if neither of these work – prisoners. The congealing of agential subjects – whether human or non-human – into static, pragmatically useful symbols-complexes seems to leave us with two options only: we can be manipulated *by* symbols, or we can be the manipulators *of* symbols. Both options would seem to lead to an isolated, disenchanted and potentially exploited mode of existence.

Symbols in Society: Towards a Renewal of Symbolic Co-creation

In short, the reality which symbolic reference allows us to inhabit brings great richness to our lives, but also terrible risks. For Whitehead, symbolism is both 'the cause of progress and the cause of error' (S 59). There is great power for the higher animals in being able to define 'those distant features in the immediate world by which their future lives are to be determined. But this faculty is not infallible; and the risks are commensurate with its importance' (S 59). As the evolution of humanity enables greater individual freedom and purposiveness, there is an accompanying increase in the risk of social fragmentation. For lower organisms, causal efficacy provides what is primarily a direct, instinctive response to particular situations (although Whitehead admits that the complexity of some of these societies, such as those of some insects, points to the limited role of symbolically coordinated action that then quickly relapses into reflex action) (S 82). But human beings, on the other hand, are able to entertain a wider range of possibilities for themselves, sometimes hopeful, but often despairing. As human beings come to possess an enhanced awareness of their own individual finitude, the risk of egotism and social fragmentation also increases.

For Whitehead, one important role of symbolism is to help compensate for the new set of risks to which human society is now prone. Symbols come to function as a kind of social glue that nonetheless works in concert *with* – rather than counter to – (some degree of) free, individual expression. As Whitehead puts it, in human life 'there are individual springs of action which escape from the obligations of social conformity. In order to replace this decay of instinctive response, various intricate forms of symbolic expression of the various purposes of social life are introduced' (S 66). For Whitehead, these systems of symbolic connection and the coordinated actions they make possible can sometimes congeal into a set of mere reflexes, contributing to the atrophy of culture through loss of novelty and 'zest'. Conversely, the rapid overturning of a certain system of symbols can throw a society into chaos. However, symbolism at its best enacts a harmonised co-ordination of action and feeling that also elicits an ongoing process of critical reflection on the reasons and purposes behind those symbols. Used in the right way, symbols can invite enquiry into their deeper meaning and justification within any given situation. This can awaken us from our torpid state of reflexive somnambulance into a more living, creative dialogue with the meaning of our symbols and thus our world.

Some helpful context for this enquiry might be found in Whitehead's more general, metaphysical reflections on the nature of evil versus progress in *Process and Reality*. Whitehead's rather curious definition of evil in this text seems, on the surface anyway, to be quite different from this term's usual association with simple moral failing. For Whitehead, the foundation of evil is built into his understanding of the atomic, eventive structure of reality – it lies in what he calls 'perpetual perishing'. The basic problem is that the more intensity and vivid beauty a given thread of occasions can marshal, the less likely it is that this thread will be able to sustain its unique life-structure through time. In Whitehead's philosophy this functions as the source

From Manipulation to Co-creation 181

of something analogous to existential anxiety: 'The world is thus faced by the paradox that, at least in its higher actualities, it craves for novelty and yet is haunted by terror at the loss of the past' (PR 340). This is to say that the greater the complexity and richness achieved, the greater is the toll that the natural tendency towards mutual obstruction will take. The higher forms of life thus make a special wager with the metaphysical conditions of reality – they take the risk of assembling something more significant, but then must also accept the condition of heightened fragility. Within this context, both outwardly destructive and self-destructive tendencies have a much higher probability of emerging.

A more straightforward account of the above would highlight the way our uniquely human capacity to envision our individual future lives leaves us haunted by the fear of death – either our own, or of those whom we love. It also leaves our species prone to a unique manifestation of egotism capable of a kind of destruction previously unheard of on this planet – the ability to degrade, possibly beyond repair, the very ecological matrix which sustains *all* planetary life. As our capacities for disenchanted melancholy and destructive egotism each come to reinforce one another, various forms of traditional religious symbolism have reasserted themselves in response, sometimes in fairly extreme ways. Fundamentalist religion of different kinds aims at addressing what is perceived to be the root of these contemporary crises by challenging modernity's liberal assertions of freedom, individualism and rationality. Reactionary religious systems function by projecting onto a transcendent sphere a reified symbolism aimed at maintaining conformity of human values and actions 'from above'. Though I am sure that many do experience fundamentalist religion as a kind of reprieve from the terrors of modern life, it constitutes neither a sound nor a sustainable solution to our collective problems. This is because it essentially repeats the trend (though in a different way from modern technoscience) of denying our unique, human destiny as *symbol-makers*. The result (in not all, but in most cases) is that it actually comes to reinforce some of the worst reflexive habits with which humanity has come to engage not just its environment, but also those perceived to be of a different religio-cultural 'tribe' (think contemporary Islamophobia).

So what is the way beyond the first option of retreating behind the projected transcendent symbols of traditional religion, and the second option of allowing our fear and insecurity to further mobilise our egoistic manipulations (either on behalf of ourselves as individuals or on behalf of our species as a whole)? Perhaps Whitehead points to a way forward in the following passage from *Process and Reality*:

> Yet the culminating fact of conscious, rational life refuses to conceive itself as a transient enjoyment, transiently useful . . . But just as the physical feelings are haunted by the vague insistence of causality, so the higher intellectual feelings are haunted by the vague insistence of another order, where there is no unrest, no travel, no shipwreck: 'There shall be no more sea'. (PR 340)

Whitehead continues this line of thought in the following paragraph:

> The process of time veils that past below distinctive feeling. There is a unison of becoming among things of the present. Why should there not be novelty without loss of this direct unison of immediacy among things? (PR 340)

In this last sentence Whitehead both clearly states the problem and hints at a way beyond it: our immersion in causal efficacy is what offers an intuitive, pre-rational sense of emerging from a larger past, and contributing to a larger future – in other words, of being integrally connected to a relationally encompassing world. The latter, however, contributes little to the uniquely human achievement of novel, vivid immediacy in our lives, which is possible only through the overlay of presentational immediacy. Mere presentational immediacy, however, tends to instil within us a deeply insecure sense of transience and hollowness. It enables a far more individualised and novel perspective on the world, but at the cost of imprisoning us within isolated egos, without any intuitive experience of living in an utterly relational world. But what if there was a sphere of reality in which harmonic intensity did not sacrifice unison of immediacy (and vice versa)? This, of course, is Whitehead's signal towards a unique conception of divinity – one whose unconditional absorption of the world (God has no negative prehensions) culminates in an everlasting divine vision that fuses limitless possibility with actual tragic beauty. Moreover, it is this divinity that conditions and releases each new event into its own unique iteration of becoming.

In these reflections, perhaps we can hear from Whitehead a call to draw from a deeper source for the courage and inspiration necessary to take up the challenge of a kind of symbol-making (and symbol-revision) conducive not just to the flourishing of our own life, but to the life of the entire bio-cultural collective of which we are a part. This 'source' would empower a form of universal connection and belongingness that at the same time would release each of us into our own unique creativity and agency. Whitehead seems to be searching for a symbol (or set of symbols) to convey this very possibility in his reflections at the end of *Adventures of Ideas*. While the symbol of 'love' comes close to capturing what he is after, he ultimately finds it too narrow in scope. For love is most commonly expressed as love of the *particular*, whereas what he is after is something quite different – namely, love of the *universal*. While Whitehead does not draw quite as severe a distinction between these two as Henri Bergson does in his late work on ethics (see *Two Sources of Morality and Religion*), like Bergson he understands that this is a difference in kind, not just degree.[4] He finally arrives at the symbol of 'peace' as more conducive to the broader meaning he is seeking. As a way of gesturing towards a kind of limited answer to the question I posed at the beginning of this chapter, I offer the following extended quote from Whitehead's *Adventures of Ideas*.

> We are in a way seeking for the notion of a Harmony of Harmonies, which shall bind together the other four qualities [Truth, Beauty, Adventure, Art] so as to exclude from our notion of civilization the restless egotism with which they have often in fact been pursued . . . I choose the term 'Peace' for that Harmony of Harmonies which calms destructive turbulence . . . It is not hope for the future, nor is it an interest in present details. It is a broadening of feeling due to the emergence of some metaphysical insight, unverbalized yet momentous in its coordination of values. Its first effect is the removal of acquisitive feeling arising from the Soul's preoccupation with itself. Thus Peace carries with it a surpassing of personality. There is an inversion of relative values. It is primarily a trust in the efficacy of Beauty . . . There is thus involved a grasp of infinitude, an appeal beyond boundaries. Its emotional effect is the subsidence of turbulence which inhibits. More accurately, it preserves the springs of energy, and at the same time masters them for the avoidance of paralyzing distinctions. The trust in the self-justification of beauty introduces faith, where reason fails to reveal the details. The experience of Peace is largely beyond the control of purpose. It comes as a gift. (AI 285)

As practically lucid as it is mystically charged, this invocation seeks a way beyond what are most commonly experienced to be life's mutually inhibiting features. Our springs of creative energy can actually be focused in a new way, rather than blunted, by expanding our awareness beyond our individuality. This passage points to an intuitive trust in the rhythmic unfolding of life that takes us beyond our usual fear of loss. It does not change the fact that what we have in this moment will at some point have to be let go of – perpetual perishing does not cease to be one of the metaphysical conditions of the world (from our creaturely perspective, anyway). But life's wondrous arc – in its singularising, spiralling motions – will emerge again and again, in countless times and places. There is a deep and omnipresent 'place' from which we can know, or at least intuit, that we are never ultimately cut off from this 'everlasting' (Whitehead's temporal description of God's 'consequent nature') spring of life's ongoing singularisation. Perhaps one of the lessons here is that if our destiny as symbol-makers on behalf of this planetary bio-collective is truly to be embraced – if it is not to fall back into a nihilistic melancholy or a ruthless egotism – it must in some sense be received as a gift from the universe. There is a strangely paradoxical logic here of having to receptively open ourselves to the very source for our most original, creative actions – one with which perhaps artists and mystics are more intimate than philosophers. And while this certainly does not offer anything like a straightforward formula for success, I do believe it might help nourish us for the ethical and ecological 'work of the world' we have before us.

Notes

1 Latour, *We Have Never Been Modern*.
2 Before shifting into the main body of the chapter, it may be beneficial to explain the initial genesis of these ideas, although the following will be primarily of interest to scholars of Henri Bergson. This project began as a comparative analysis of Bergson and Whitehead on the evolutionary emergence of perception and meaning-making in the human organism, especially tracing some strong similarities in the way their respective insights could be – and indeed were, by them – applied to contemporary ethical problems. Upon further research, however, it became clear to me that Bergson had little direct interest in the dynamics of symbolism per se, and thus did not serve as an altogether productive interlocutor with Whitehead on these concepts. Nevertheless, anyone familiar with Bergson's philosophy will very quickly pick up on the Bergsonian themes that I am drawing out of Whitehead's work. When I discuss the evolutionary origins of perception in terms of a pragmatic 'filter' and the way it has led to the fallacious epistemology of modern science, these insights are in some ways more quintessentially Bergsonian than Whiteheadian. Similarly, my attempt to find a 'third' approach to symbolism that combines novel individuality (made possible by 'presentational immediacy') with a renewed

awareness of our immersion in a larger flow of becoming, has strong correlations with Bergson's project – especially as developed in his late work, *The Two Sources of Morality and Religion* – of forging a third path beyond the dominant options of 'instinct' and 'intelligence' that he calls 'intuition' – also identified with what he calls 'open religion'. It is my hope that these powerful resonances in their thought might be further developed – either by myself or others – in future scholarship.
3   Abram, *The Spell of the Sensuous*.
4   Bergson, *Two Sources of Morality and Religion*.

Bibliography

Abram, David, *The Spell of the Sensuous: Perception and Language in a More-Than-Human World* (New York: Knopf Doubleday, 1997).

Bergson, Henri, *Two Sources of Morality and Religion*, trans. R. Ashley Audra and Cloudesley Brereton (Garden City: Doubleday Anchor, 1935).

Latour, Bruno, *We Have Never Been Modern*, trans. Catherine Porter (Cambridge, MA: Harvard University Press, 1993).

Whitehead, Alfred North, *Adventures of Ideas* (New York: The Free Press, [1933] 1967).

Whitehead, Alfred North, *Modes of Thought* (New York: The Free Press, [1938] 1966).

Whitehead, Alfred North, *Process and Reality: An Essay in Cosmology*, corrected edition, ed. David Ray Griffin and Donald W. Sherburne (New York: The Free Press, [1929] 1978).

Whitehead, Alfred North, *Symbolism: Its Meaning and Effect* (New York: Fordham University Press, [1927] 1985).

10

On Symbols, Propositions and Idiocies: Towards a Slow Technoscience

A. J. NOCEK

> It is the first step in sociological wisdom, to recognize that the major advances in civilization are processes which all but wreck the societies in which they occur: – like unto an arrow in the hand of a child. (Alfred North Whitehead, *Symbolism: Its Meaning and Effect*)

Alfred North Whitehead's Barbour-Page lectures, published as *Symbolism: Its Meaning and Effect* (1927), are often overshadowed by his much more substantial Gifford lectures (1927–8), published as *Process and Reality*. This should come as no surprise, given that *Symbolism* offers a less thorough treatment of the topic than *Process and Reality*. For this reason, *Symbolism* has been largely forgotten. What is often overlooked, however, is that *Symbolism* contains one of the few clear instances where Whitehead speaks about symbolism's importance for the maintenance and destruction of 'political societies'. For example, he talks about the significance of symbolism in the English revolutions in the seventeenth century and the American Revolution in the eighteenth century (S 77–8). He also underscores the importance of balancing novel revisions of and reverence for the symbolic codes of a political society; without this, the society will decay 'from anarchy, or from the slow atrophy of a life stifled by useless shadows' (S 88).

This chapter takes seriously the idea that Whitehead's work on symbolism in the Barbour-Page lectures, and elsewhere, is still relevant for understanding the political-economic conditions of our twenty-first-century epoch. To do so, I draw on a set of contemporary practices whose use of symbolism eclipses new patterns of value: today's technoscientific industry. I contend that Whitehead's work on symbolism elucidates how technoscientific production has been captured by a system of political and economic meanings (neoliberalism) which tends to disqualify modes of resistance. I draw heavily on Isabelle Stengers' recent plea for a 'slow science' in the face of fast and competitive technoscience in order to expose how it is that we are in dire need of new forms of symbolism in today's scientific knowledge economy. Along the way I consider how Whitehead's notion of the 'proposition' in *Process and Reality* makes a key intervention in this discussion and reinforces the importance of symbolism in the culture of twenty-first-century technoscience. Ultimately, the chapter contends that technoscience requires new propositions for feeling its products and practices outside of neoliberalised symbolic codes. The figure of the 'Idiot scientist' provides the propositional lure for feeling technoscience differently.

The Idiot Scientist

In 2011, Isabelle Stengers gave the Willy Calewaert Leerstoel lecture at the Vrije Universiteit Brussel titled '"Another science is possible!" A plea for slow science'. Right away, Stengers expresses her gratitude for being invited to give the lecture in spite of her suspicion that many people think that she is a 'troublemaker'. And a troublemaker is precisely what she is, or in any case I shall argue that she is. But the question I am concerned with is, 'For whom is she a troublemaker?' Stengers wastes no time by taking sides with another troublemaker in the Belgian scientific community. In no uncertain terms, she explains that Barbara Van Dyck is a scientist who was 'sacked' by the Katholieke Universiteit Leuven 'for having publicly explained and endorsed the action against a genetically modified potatoes field in Wetteren'.[1]

According to Stengers, Van Dyck's act of 'civil disobedience' was met with a strong response from the Leuven authorities: her immediate dismissal before any penal decision had been reached. The University's actions were meant to send an equally strong message to the scientific community: namely, that Van Dyck's form of resistance was an 'act of violence' (sic) against scientific research itself. But to Stengers' mind the real message that sacking Barbara Van Dyck sent is that the primordial democratic right of civil disobedience is an 'act of violence' against fast and competitive science, or what has become the new gold standard of scientific research.

To say that there exist strong ties between scientific research and industry is not new of course. For decades, historians, sociologists and philosophers of science and technology have recognised that there is little to no barrier separating the work of scientists from developments in industry. In the US, for example, many scholars attribute key legislation, such as the Bayh-Dole Act of 1980, to the emergence of the 'entrepreneur scientist';[2] a model of 'scientific subjectivity' that has been welcomed with open arms in the European context as well.[3] But what was new about the Van Dyck affair, and is worthy of more attention, was how the Leuven authorities vilified her actions.

Crucially, the Rectorate of Leuven denounced Van Dyck's form of resistance because it was directed against experiments performed by 'colleague scientists'. Stengers is quick to note that this justification makes it seem as if the experiments on genetically modified potatoes were produced under scientific conditions that are essentially neutral, and only incidentally industry bound. In other words, the Rectorate's remarks pit Van Dyck's actions against the tried and true methods of scientific experimentation. The real problem with this justification, according to Stengers, is that Van Dyck was not acting from her position as a 'scientist', that is, as someone whose job it is to make objections in order to test the robustness and reliability of scientific claims, but, rather, from her position as a *citizen*. In other words, what is so troubling about this case is that Van Dyck's objection was based on knowledge that any other citizen could have acquired; but as it turns out, the concerned citizen is a subject position from which she is not allowed to object.

As Stengers succinctly puts it, '[h]er dismissal . . . means that the Leuven university feels entitled to control its workers' whole life and not only what they do at work, that is, in this case, as researcher'.[4] The university resembles 'the medieval corporation', she goes on to claim, which means that 'the member of a corporation was indeed not a citizen but a member of a body, with no independent life'. And as a part of the university qua medieval corporation, 'even when they act as citizens, researchers will now have to respect the interest of their corporation'.[5]

While Stengers does not take her talk in this direction, it should be noted that the Van Dyck affair also vividly demonstrates what Michael Hardt and Antonio Negri, along with Maurizio Lazzarato, Paolo Virno and others describe as the transition from capitalism's 'formal subsumption' to its 'real subsumption' of labour.[6] It is not simply that subjects labour during working hours, but they are labouring in all other aspects of their lives – in their cognitive and affective lives as well. This new form of labour is what Hardt and Negri call 'immaterial labor'[7] and coincides with the emergence of what has been variously termed the 'information economy' or the 'post-Fordist economy'.[8] In this register, Van Dyck's workday does not end when she leaves her lab, when she bikes home, or when she retires for the evening; in all of these activities she is labouring. This, it seems to me, is one of the crucial takeaways of Van Dyck's dismissal: namely, that she is not 'free' in her off-hours to place the claims of scientists into contexts that may frustrate competitive science, that make genetically modified organisms (GMOs) accountable to the many environments that they affect. Van Dyck resisted the general conditions of work under neoliberal capitalism, and in this way her civil disobedience makes trouble. She is a troublemaker, or maybe even an 'Idiot'.

I use the term 'Idiot' here not in the pejorative sense, but in the way that Gilles Deleuze invokes it in *Difference and Repetition* to praise Idiocy in 'the Russian manner'.[9] In Deleuze's view, the Idiot in Dostoyevsky's *Notes from the Underground* protests 'what everybody knows' and what passes for 'common sense'.[10] The 'underground man', Deleuze explains, 'recognizes himself no more in the subjective presuppositions of a natural capacity for thought than in the objective presuppositions of a culture of the times, and lacks the compass with which to make a circle'.[11] Similarly, for Stengers the Idiot 'is the one who always slows the others down, who resists the consensual way in which the situation is presented and in which emergencies mobilize thought or action'.[12] Furthermore, 'the idiot demands that we slow down, that we don't consider ourselves authorized to believe we possess the meaning of what we know'.[13] For Deleuze, as for Stengers, Idiocy refuses to authorise a hidden source of knowledge that transcends the situation; it resists the temptation to place faith in an arbiter who can validate and disqualify others' claims. The Russian Idiot approaches a situation without presumption – or a 'dogmatic image of thought' – and thus possesses no mastery over the rules that govern it.[14]

Stengers explains that it is Deleuze's Idiot persona, and not Kant's 'citizen of the cosmos', that is behind her cosmopolitical proposal in her multi-volume *Cosmopolitics*. Indeed, that her cosmopolitics might be confused with Kantian cosmopolitanism, a state of affairs in which everyone might be a member of a world citizenry, a 'citizen of the cosmos', worries Stengers: 'This is where the proposal is open to misunderstanding', she writes, and is susceptible to being interpreted as a politics that 'aim[s] at allowing a "cosmos", a "good common world" to exist – while the idea is precisely to slow down the construction of this common world, to create a space for hesitation regarding what it means to say "good"'.[15] The cosmopolitical refuses to give a definition of what is held in common and by whom. On the contrary, it calls upon us to abstain from answering on behalf of others and knowing in advance who those others will be. This is why '[t]he cosmos', she continues,

> must therefore be distinguished here from any particular cosmos, or world, as a particular tradition may conceive of it. Nor does it refer to a project designed to encompass them all . . . In the term cosmopolitical, cosmos refers to the unknown constituted by these multiple, divergent worlds, and to the articulations of which they could eventually be capable, as opposed to the temptation of a peace intended to be final, ecumenical: a transcendent peace with the power to ask anything that diverges to recognize itself as a purely individual expression of what constitutes the point of convergence of all.[16]

In this perspective, there is a very particular cosmos of late capitalism that disqualified Van Dyck's refusal to allow fast technoscience to answer on behalf of Wetteren potatoes. That she supported the Field Liberation Movement resulted in her termination.[17] It is for this reason that Van Dyck, I am tempted to suggest, exemplifies the Idiot persona: she slowed down competitive technoscience's march forwards by entertaining the objections of those whom the market deems unworthy of making objections. In short, she dared to challenge what 'everybody knows'. But if Van Dyck is indeed an Idiot, she is a new kind of Idiot: not a literary or philosophical Idiot, but a *scientific Idiot*.[18]

Civilising the Idiot

When Stengers characterises 'slow science' she often draws on Whitehead's thought. In the lecture she gave at Vrije Universiteit Brussel, for example, she references the first sentence of the epilogue to Whitehead's 1938 lectures published as *Modes of Thought:* the 'task of the university', he insists, 'is the creation of the future, so far as rational thought, and civilized modes of appreciation, can affect the issue. The future is big with every possibility of achievement and of tragedy' (MT 171). Stengers is quick to mention that this 'future' does not reference 'progress', but 'uncertainty'; and this uncertainty is somehow bound to the promotion of 'rational thought' and 'civilized modes of appreciation'. But just what does Whitehead mean by rational thought and civilised modes of appreciation in this context, and how might they help characterise the meaning of slow science?

Stengers reminds us that what worried Whitehead about education here, and elsewhere, is the nineteenth-century 'discovery of the method of training professionals, who specialize in particular regions of thought and thereby progressively add to the sum of knowledge within their respective limitation of subject' (SMW 196). Of course, Whitehead is critiquing neither abstraction nor specialisation per se (which is why he does indeed offer an alternative to the Kantian critical tradition);[19] after all, we 'cannot think without abstractions' (SMW 59), and speculative philosophy requires its own methods of abstraction. Rather, what concerns Whitehead are the devastating effects of modern professionalism: it produces 'minds in a groove', in which professionals 'live in contemplating a given set of abstractions. The groove prevents straying across country, and the abstraction abstracts from something to which no further attention is paid' (SMW 197). The risks associated with this 'groove' are profound:

> The directive force of reason is weakened. The leading intellects lack balance. They see this set of circumstances, or that set; but not both sets together. The task of coordination is left to those who lack either the force or the character to succeed in some definite career. (SMW 197)

Whitehead warns us that this nearly ubiquitous habit of modern thought commits what he calls the 'fallacy of misplaced concreteness', or the 'accidental error of mistaking the abstract for the concrete' (SMW 51). One of the central tasks of his speculative project is to devise rational and civilised modes of thought that resist this tendency, that put 'all we know of nature . . . in the same boat, to sink or swim together' (CN 148).[20] Whitehead was, however, particularly worried about the vulnerability of scientific education to this fallacy: 'the modern chemist', he explains,

> is likely to be weak in zoology, weaker still in his knowledge of the Elizabethan drama, and completely ignorant of the principles of rhythm in English versification . . . Of course I am speaking of general tendencies; for chemists are no worse than engineers, or mathematicians . . . Effective knowledge is professionalised knowledge, supported by a restricted acquaintance with useful subjects subservient to it. (SMW 197)

In short, professional knowledge prevents 'civilized knowledge' from taking hold, where the latter indexes knowledge that 'surveys the world with some large generality of understanding' (MT 4). The professional's 'restricted acquaintance with useful subjects subservient to it' fundamentally lacks the balance that civilised modes of thought cultivate.

Much of what propels this imbalance is the scientist's belief in the 'finite fact': it is a 'myth', Whitehead insists, which neglects to take into consideration 'the environmental coordination requisite for its existence. This environment, thus coordinated, is the whole universe in its perspective to the fact' (MT 9). Professional knowledge, which Whitehead also calls 'effective knowledge', is incapable of addressing itself to its environment, 'which, in its totality we are unable to define' (MT 10). That we can never know this environment 'in its totality' is crucial for Whitehead, inasmuch as it indicates that there is always an uncertainty, or an 'interstice', in our understanding of the relations into which a fact enters.[21] A 'fact' has an incalculable number of environments and is prehended by them in an incalculable number of ways. This uncertainty, which is the same uncertainty, I believe, that Whitehead champions in university education, is what is denied to the professional. Whitehead laments that

> the rate of progress in such an individual human being [the professional], of ordinary length of life, will be called upon to face novel situations which find no parallel in the past. The fixed person for the fixed duties, who in older societies was such a godsend, in the future will be a public danger. (SMW 196)

What concerns Whitehead is the rate at which knowledge is professionalised seems to correspond to the rate at which its environmental coordination is neglected. How, for example, an abstraction is supported or undermined by the way in which it is incorporated into other environments is not of interest to the professional. Professionalism fails to account for the fact 'that every society requires a wider social environment', which 'leads to the distinction that a society may be more or less "stabilized" in reference to certain sorts of changes in that environment' (PR 100). It is the professional's disregard for this 'ecological fragility' that makes him or her a 'public danger' in Whitehead's view.

Arguably, professionalism has only grown worse since the early twentieth century. Although he could not have foreseen the extent to which the sciences would professionalise, Whitehead's early intuitions about professionalism seem to anticipate the bewildering number of scientific specialisations today. Nanotechnology, synthetic biology, bioinformatics, genetic engineering and so on – these are all technosciences that did not exist in Whitehead's day, but they are growing and adding subfields at an unprecedented rate. Add to this the profound disconnect between scientific and humanistic education in the university today, and it seems that Whitehead's worst fears have been realised.[22]

It could be argued that I am overstating the case a bit. For example, anyone familiar with twenty-first-century technoscience knows that collaboration is the name of the game. Quantum physicists are collaborating with engineers in nanotechnology;[23] genomics researchers and computer scientists are producing stunning data visualisations together;[24] electrical engineers are teaming up with synthetic biologists to create biological circuits;[25] designers and architects are even weighing in on the 'redesign' of living systems in synthetic biology;[26] and humanities centres around the United States (as elsewhere) are beginning to take notice of developments in the technosciences.[27] So if these collaborations, and others like them, are now the rule rather than the exception, aren't Whitehead's concerns about the future of education misguided? Or in any case, aren't we already championing the kind of cross-pollination that invigorated Whitehead's thought and is key to cultivating civilised modes of appreciation?

It is at this point that Stengers' insights on the culture of scientific knowledge in the twenty-first century are crucial. We must not forget, she insists, that the 'reliability' of a scientific claim does not depend upon its 'objectivity',[28] but springs from 'competent colleagues' whose job it is to raise objections, to ensure that the claim has been submitted to demanding tests.[29] It is the 'shared concern' among scientists that a claim is 'reliable' in the face of shifting environments that justifies its exposure to harsh objections from colleagues.

What characterises today's climate of technoscience, however, is the loss of passionate objectors who guarantee the reliability of scientific claims: scientists, she argues, are now bound by the already settled interests of industry.[30] In other words, the reliability of the claims and products of competitive technoscience are relative to the 'purified' environments from which objections to them have been made,[31] and these environments are ones that now serve market interests. Thus, today's 'collaborations' in technoscience do not have the same meaning as Whitehead's civilised modes of appreciation. If anything, these collaborations express the fact that 'minds [are] in a groove' more so now than ever before. Indeed, there is a sweeping reduction of all interests to a common measure: market rationalism. Scientific minds are in a *neoliberal* groove.[32]

What this means is that objections that are made from outside of this common measure (if it is even possible to secure such a position today) are disqualified. They are 'messy' complications that must be silenced.[33] Van Dyck is just such a messy complication: she posed an objection from a space where objections are not tolerated. For this reason, she has been turned into an 'enemy' of science. What worries Stengers is that by abstracting from this 'mess' fast and competitive technoscience lacks the balance of 'civilized knowledge', and this puts scientific reliability at risk. The claims and products of technoscience are circumscribed by one sphere of value. Products developed in synthetic biology labs, such as novel enzymes for biofuels, are made with an eye to competitive markets and with little regard for how they may have negative effects on the global food supply. To raise an objection about the global food supply with a working synthetic biologist is to slow things down; it is to be an Idiot.[34]

The Idiot's Proposition

What the above discussion makes clear is that the products of technoscience have an incredibly limited sphere of meaning for professionals who work in the field; their value rests on the markets they generate. In short, their meaning is circumscribed by the symbolic code(s) of neoliberal ideology. Given this, the central provocation of this chapter is that twenty-first-century technoscience is in dire need of new kinds of symbolism. If Whitehead's 1927 lectures at the University of Virginia still resonate for us today, as I claim they should, it is not only because they showcase the absolutely central role that symbolic reference plays in human and non-human perception, but also because they are a crucial touchstone for understanding how it is possible to generate meanings that are not determined by neoliberalised symbolic codes.

Whitehead is careful to point out in his lectures that not every epoch values symbolism and that some epochs have even tried to get rid of it altogether. 'With the Reformation', for example, '[m]en tried to dispense symbols as "fond things, vainly invented", and concentrated on their direct apprehension of the ultimate facts' (S 1). Indeed, '[t]he very fact that it can be acquired in one epoch and discarded in another epoch testifies to its superficial nature' (S 1). Arguably, one of Whitehead's central concerns in these lectures is to dispel the myth that symbolism is a mere accessory to human behaviour and to bring its constitutive role in human thought and civilisation into full view.

'Symbolism is no mere idle fancy or corrupt degeneration', he insists, 'it is inherent in the very texture of human life . . . you reduce the functions of your government to their utmost simplicity, yet symbolism remains' (S 61–2).

It is with this in mind that I examine how today's technoscientific culture can be framed in terms of Whitehead's theory of symbolism. This exercise is useful not simply because it gives a technoscientific problem a new vocabulary, but also because it sharpens our understanding of where, in Whitehead's view of 'civilized science', alternative meanings come from; how it is possible to entertain new relevancies; and under what conditions their 'truth' matters. If there is indeed a scarcity in the meanings that we give to GMOs, synthetic cells, nano-parts and the host of other technoscientific creatures, then it is of the utmost importance that we investigate how it is possible to cultivate new meanings in Whitehead's view, even when – or rather, especially when – they are regarded as 'Idiotic'.

To do so requires some preliminary remarks on Whitehead's understanding of symbolic reference. For Whitehead, symbolic reference has a deeply pragmatic function, inasmuch as it connects an object of present perception to past and future experiences. In other words, symbolic reference allows there to be an 'emotional' prolongation of a thing in perception. Thus, Whitehead's notion of the 'symbolic' could not be further from Lacan's 'death of the thing':[35] symbolic reference, on the contrary, is the 'synthetic element contributed by the nature of the percipient' that enables 'some components of its experience [to] elicit consciousness, beliefs, emotions, and usages, respecting other components of its experience' (S 8). Symbolic synthesis is not the exclusive privilege of human animals, however. Whitehead recognises that meaning-making is indispensable for organismic survival in general – for example, 'that rustle in the bushes is a predator'. So although Whitehead's focus on symbolism in these lectures might suggest a latent humanism, especially insofar as human language is the centrepiece of his investigations, it is crucial to keep in mind that language is only a highly developed species of the much more general activity of symbolisation. In the Gifford lectures, published as *Process and Reality*, Whitehead explains that:

> [l]anguage is the example of symbolism which most naturally presents itself for consideration of the uses of symbolism. Its somewhat artificial character makes the various constitutive elements in symbolism to be the more evident. For the sake of simplicity, only spoken language will be considered here. (PR 182)

But no matter how simple or complex symbolic references are they are always the synthesis of two more primary modes of pure perception: 'presentational immediacy' and 'causal efficacy'. Where presentational immediacy yields a pure phenomenology of the world's simultaneity and objects vividly appear as sense-data that have no past or future (see S 24; PR 123), causal efficacy indexes a world in which objects enter into the constitution of each other – that is to say, past experiences are actually *in* present experiences that will be incorporated into future experiences (see S 41, 44, 51; PR 170).[36]

It is crucial to recognise here that presentational immediacy and causal efficacy are infallible in themselves. Presentational immediacy *just is* the immediate and vivid presentation of sense-data; and causal efficacy *just is* the constitution of a percipient from the influx of past experiences (see S 81). Fallibility arises in the synthetic act of symbolic reference. Whitehead makes this point at a few crucial places in the lectures and alludes to the fact that symbolic reference always risks being wrong (see S 19) insofar as it arises from the false attribution of causal efficacy (S 81). The idea is that in the act of blending presentational immediacy and casual efficacy a mistake has been made – the rustling in the bushes, for example, was not actually a predator, it was only the wind.

Symbolic reference therefore introduces into experience the possibility of error, but it also introduces the possibility of learning from experience.[37] This is why error must not be judged too harshly: indeed, Whitehead will go so far as to claim that error is what accounts for 'imaginative freedom' and 'mental progress' (see S 19). In *Process and Reality*, Whitehead reinforces this view when he tells us that the 'evolutionary use of intelligence is that it enables the individual to profit by error without being slaughtered by it' (PR 168). Thus, error is not always detrimental to experience; in many instances it is even advantageous: the possibility of error is what allows minds to escape what is simply given and imagine possibilities not already exemplified in experience.

The reason I have drawn attention to 'error' is that it underscores the fact that there is always an element of uncertainty that accompanies symbolic reference. Whitehead is of course sensitive to the fact that not all organisms experience the uncertainty of symbolism, that is, *feel* that what is entertained might not be confirmed by the actual world – that it is only a 'theory'.[38] But what metaphysical status does Whitehead give these possible and not necessarily actual meanings, which are often useful inasmuch as they lure us into feeling the given in new ways? In other words, what kinds of entities are these 'potential meanings', which always risk being wrong, but very clearly pave the way for the 'imaginative freedom' that Whitehead champions?

It is precisely in this area of conceptual development where the Barbour-Page lectures are less developed than the Gifford lectures. In *Process and Reality* Whitehead carefully dissects how living systems are able to feel sense-data in a 'potential relation' that have yet to be verified. He will call these feelings 'propositional'. Without exhaustively recounting Whitehead's theory of propositional feelings, suffice it to say that the proposition is neither true nor false in itself, but it is a real 'hybrid entity' (more on this later)[39] which 'lures us into feeling' a world that 'might be'. More precisely, 'the primary mode of realization of a proposition in an actual entity is not by judgment, but by entertainment. A proposition is entertained when it is admitted into feeling' (PR 188). The proposition is thus the 'could be' of an eternal object, or predicate, with a subject for a concrescing occasion. It is the 'potential togetherness' of subject and predicate, and not its actual realisation, which defines a proposition: 'the germaneness of a certain set of eternal objects to a certain set of actual entities' (PR 188).

While both propositions and eternal objects have 'potentiality' in common, Whitehead is careful to distinguish the two. Eternal objects are what he calls 'pure potentials' (PR 22), which means that they do not reference any particular actuality or epoch. Propositions, on the contrary, are impure, in the sense that they always involve a particular set of subjects, in a particular epoch, that are put into a potential relation with a predicate. These subjects are 'reduced to the status of food for a possibility' (PR 258). And if the 'presupposed logical subjects' do not exist 'in the actual world of some actual entity . . . the proposition does not exist for that actual entity . . . The proposition itself awaits its logical subjects. Thus, propositions grow with the creative advance of the world' (PR 188).

Crucially, Whitehead's reflections on the proposition sharpen our understanding of how meanings do not have to conform to the actual conditions of the world to be useful. To entertain a meaning for a symbol is to place a logical subject in a potential predicative pattern. 'Truth talk' is irrelevant at this stage. That a concrescing occasion feels what is given in a pattern that turns out to be 'non-conformal' to the actual world does not mean that the feeling is 'merely wrong, and therefore worse than useless' (PR 178). Rather than treat propositions as 'materials for judgment', as he did in his co-authored work with Bertrand Russell,[40] Whitehead now contends that propositions are fully real, hybrid entities that are among the 'eight categories of existence', irrespective of their 'truth value'. 'Its own truth, or its own falsity, is no business of a proposition', Whitehead claims. 'That question', he continues, 'concerns only a subject entertaining a propositional feeling' (PR 258). Propositions open up alternative ways of feeling the world instead of merely reflecting upon or representing it. At stake are the kinds of feelings that propositions make possible rather than their truth-value. This is why Whitehead famously says that 'it is more important that a proposition be interesting than that it be true' (PR 259).

Conclusion

The whole point of venturing into Whitehead's theory of symbols and propositions is to say something about today's technoscientific culture; and in particular, about the culture that vilifies scientists who oppose the meanings that 'everyone knows'. In this concluding section of the chapter, I want to suggest that Whitehead's work on symbolism brings the foreclosure of new propositional feelings in the Van Dyck affair into full view. In other words, what is denied to Van Dyck is the possibility of entertaining alternative meanings to GMO production. What she is unable to do – or rather, she tries to do and is sacked for – is entertain a proposition that opens up an 'interstice' in the well-worn meanings that GMOs have been given by industry. In today's technoscientific culture, the process of abstraction or propositional entertainment, which generates the space for possibility and error, is often foreclosed to scientists. To place GMOs in a predicative pattern that would lure us into feeling the effects of genetic modification in terms of the Wetteren farmlands places the values of industry at risk. The wealth of causal efficacy is eclipsed by competitive technoscience and those scientists who have not fallen into the neoliberal groove must be silenced. Their propositions are regarded as 'false' and easily disqualifiable since they do not conform to the 'true' conditions of the world. Market values are representative of those 'true' conditions and are supposed to save us from superstitious beliefs.[41]

But what Whitehead's work also makes abundantly clear is that these propositional 'lures' are not 'materials for judgment', as both the history of logic and neoliberal capitalism want to insist; they are entities in their own right that lure us into feeling a situation differently. They are lures that expose thought to the wealth of possibilities issuing from the efficacious environment, but which have been denied to us by late capitalism's firm grip on scientific imagination. They are lures that make possible 'civilized modes of appreciation'. In this way, the verity of Van Dyck's claim is far less important than the fact that she submits GMOs to new modes of entertainment. Thus, when Stengers advocates 'slow science' in her work, I would argue that she is also devising the conditions for technoscientific products to become 'logical subjects' for new predicative patterns. In other words, her plea for slow science is also a plea for constructing 'lures for feeling' technoscience differently.

One of the central concerns of this chapter has been to investigate how new meanings are generated for technoscience under conditions of neoliberal capitalism. Whitehead has done important work for us insofar as his theory of propositional lures for feeling articulates how new modes of entertainment come into being. But how are these lures cultivated? This, it seems to me, is the question that ties Whitehead's metaphysical speculations about symbols and propositions to our present epoch – a bond that he requires of every metaphysical abstraction.[42] My argument is that Idiocy is what creates space for these imaginative lures: it slows down the pace of technoscientific production by refusing to give it the meaning that 'everybody knows'. The Idiot is the one who just cannot seem to get on board. The scientific Idiot is in no way anti-technoscience, however. In fact, she is just the opposite: her Idiocy promotes the 'reliability' of technoscientific claims and production by levelling objections to them from outside of the purified environments of industry. Idiocy therefore dares to cultivate 'civilised' lures for feeling the practices and products of technoscience when they would otherwise be wholly resistant to them.

It is important to keep in mind that Idiocy is itself a speculative proposition, and it is hard to know whether it was actualised by Van Dyck or not. More important than this, I believe, is that Idiocy is a proposition that lures us into different ways of inhabiting the technoscientific world, of situating ourselves within 'what everybody knows', when we seem to be lacking such lures today. That we have somehow run out of ways to entertain the future, that neoliberalism has somehow sapped all of the hope from us, is commonly heard among neo-Marxists today.[43] Put in a Whiteheadian key, we might say that neoliberalism ensures that there are no new patterns for logical subjects to entertain, that propositions *do not* 'grow with the creative advance of the world'. That Whitehead would have something to say to neo-Marxists is less surprising than it used to be.[44] As such, I think it is worth entertaining the proposition that scientific Idiocy could speak to the urgency felt among scholars of all stripes to entertain futures that are not circumscribed by one symbolic code. I think the closing lines of Whitehead's Barbour-Page lectures capture the sentiment best:

> The art of a free society consists first in the maintenance of the symbolic code; and secondly in the fearlessness of revision, to secure that the code serves those purposes which satisfy an enlightened reason. Those societies which cannot combine reverence to their symbols with freedom of revision, must ultimately decay either from anarchy, or from the slow atrophy of a life stifled by useless shadows. (S 88)

Notes

1 Stengers, 'A plea for slow science', 2.
2 See Cooper, *Life as Surplus*; Rajan, *Biocapital: The Constitution of Postgenomic Life*; Mitchell, *Bioart and the Vitality of Media*; and Daston and Galison, *Objectivity*.
3 Pellizzoni and Ylönen, *Neoliberalism and Technoscience*.
4 Stengers, 'A plea for slow science', 3.
5 Stengers, 'A plea for slow science', 3.
6 Michael Hardt and Antonio Negri argue that 'real subsumption' references the full interiorisation of social relations to capital (Hardt and Negri, *Empire*, 225, 229). These new conditions of labour – for example cognitive, affective and immaterial – are elaborated by the autonomia and post-autonomia tradition of Marxism, particularly in work by Maurizio Lazzarato, Paulo Virno, Franco 'Bifo' Berardi and others. Recently, Steven Shaviro has argued that it is more accurate to theorise the passage from formal to real subsumption as a 'tendency', rather than as inevitable; there is no teleological closure. Thus, real subsumption is never fully accomplished but is always in the making. See Shaviro, *Post-Cinematic Affect*, 188; and Marx, *Capital*, vol. 3, 317–75.
7 The notion of 'immaterial labor' references a change in the modes of capitalist production that are tied to the emergence of a post-Fordist economy. The Italian tradition of *operaismo* links immaterial labour to Marx's notion of the 'general intellect' in the 'Fragment on machines' in the *Grundrisse*. As Paulo Virno argues, Marx foresaw that the assembly line would be relegated 'to the fringes', and 'abstract knowledge' would become the main productive. Automated machinery will replace manual labour. See Virno, 'General intellect'; Lazzarato, 'Immaterial labor'; and Hardt, 'Affective labor'.

8 See for example, Foucault, *The Birth of Biopolitics*; Brown, 'Neoliberalism and the end of liberal democracy'; Hardt and Negri, *Empire*, 22–41.
9 Deleuze and Guattari return to the Idiot persona in *What Is Philosophy?* and contend that it has two manifestations: the 'old' and the 'new' Idiot. The 'old' Idiot is Cartesian, and is the one 'who says "I" and sets up the cogito . . . who wants to think, and who thinks for himself, by the "natural light"' (61–2). But where the 'old idiot wanted truth . . . the new idiot wants to turn the absurd into the highest power of thought – in other words, to create . . . The new idiot will never accept the truths of History' (62–3). Also see Flaxman, *Gilles Deleuze and the Fabulation of Philosophy*, 187–8; and Rajchman, *The Deleuze Connections*, 37–8.
10 See Deleuze's critique of the 'dogmatic image of thought', which operates according to 'common sense' and 'good sense', in *Difference and Repetition*, ch. 3, 'The image of thought'. Also see Lambert, *In Search of a New Image of Thought*, for a careful and illuminating discussion of Deleuze's critique of common sense throughout his work.
11 Deleuze and Guattari, *What Is Philosophy?*, 130
12 Stengers, 'The cosmopolitical proposal', 994.
13 Stengers, 'The cosmopolitical proposal', 995.
14 In *What Is Philosophy?*, Deleuze and Guattari write that, '[t]he new idiot has no wish for indubitable truths; he will never be "resigned" to the fact that 3+2=5 and wills the absurd . . . The old idiot wanted, by himself, to account for what was or was not comprehensible, what was or was not rational, what was lost or saved; but the new idiot wants the lost, the incomprehensible, and the absurd to be restored to him' (62–3).
15 Stengers, 'The cosmopolitical proposal', 995. Stengers admits that she was unaware of Kant's cosmopolitanism when she used the concept: 'I have to plead guilty since I was unaware of Kantian usage when, in 1996, while working on the first volume of what was to become a series of seven *Cosmopolitiques*, this term imposed itself on me, so to speak. I therefore wish to emphasize that the cosmopolitical proposal, as presented here, explicitly denies any relationship with Kant or with the ancient "cosmopolitism". The "cosmos", as I hope to explain it, bears little relation to the world in which citizens of antiquity asserted themselves everywhere on their home ground, nor to an earth finally united, in which everyone is a citizen' (994). In conversation, Catherine Keller has recently suggested to me, however, that Stengers' quick dismissal of Kant's notion of the common risks making her cosmopolitical proposal politically irrelevant. Crucially, Stengers fills out her notion of 'shared relations' through what she calls 'etho-ecology', or, in other words, the idea that 'ecology must always be etho-ecology [and] there can be no relevant ecology without a correlate ethology, and . . . there is no ethology independent of a particular ecology' (Stengers, 'Introductory notes on an ecology of practices',

187). In recent years, Roberto Esposito has also weighed in on the possibility of there being a notion of 'communitas', or *bios* held in common, that does not result in thanatopolitical violence (see Esposito, *Bíos: Biopolitics and Philosophy*). For the limitations of this view, see Wolfe, *Before the Law*.
16. Stengers, 'The cosmopolitical proposal', 995.
17. See http://www.fieldliberation.org/en (last accessed March 2017).
18. In *Difference and Repetition*, Deleuze writes that, '[t]he philosopher takes the side of the idiot as though of a man without presuppositions' (130). Frida Beckman explores the conceptual persona of the literary idiot at length in her article 'The idiocy of the event'. To my knowledge, Stengers is the only one who has suggested the possibility of 'scientific idiot' (Stengers, 'Introductory notes on an ecology of practices'), although she does not clarify how this persona might differ from other idiot personae.
19. Gaskill and Nocek, 'Introduction: an adventure of thought'.
20. On 'civilized modes of thought' in Whitehead, see *Modes of Thought*, especially lecture VI.
21. See Debaise, 'The living and its environments', and Stengers, *Thinking with Whitehead*, 312–35.
22. Williams, 'Whitehead's curse?'
23. Band and Avishai, *Quantum Mechanics with Applications to Nanotechnology and Information Science*.
24. McGill, 'Molecular movies . . . coming to a lecture near you'.
25. Endy, 'Nature is designed'.
26. Ginsberg et al., *Synthetic Aesthetics*.
27. To take one example, in 2012 the Simpson Center for the Humanities at the University of Washington sponsored an initiative with the Fred Hutchinson Cancer Research Center titled 'Biological Futures in a Globalized World'. This partnership culminated in an international symposium titled 'Synthetic Biology in Question'. The point is that evidence for these kinds of collaborations now abound. SUNY Albany's initiative, the Center for Humanities, Arts and Technoscience, is another striking example of how the humanities are taking notice of developments in the sciences, and are investing in those who are willing to work on these developments.
28. Stengers' dismissal of 'objectivity' in the sciences resonates, in certain ways, with the internalist history of scientific epistemology that Lorraine Daston and Peter Galison offer in their work *Objectivity*. There, Daston and Galison argue that objectivity is an epistemic norm that emerged in the nineteenth century and has since been replaced by other values. Similarly, Stengers dismisses the hold that objectivity has over scientific epistemology, and argues that scientific reliability springs from elsewhere.
29. Stengers, 'A plea for slow science', 9.
30. See Pellizzoni and Ylönen, *Neoliberalism and Technoscience*.

31 Also see Stengers' illustrative example of the 'purified environments' of animal experimentation in her essay 'The cosmopolitical proposal'.
32 Michael Halewood has suggested to me that it might be worth modifying Whitehead's phrase 'minds in a groove' to express how we are captured by neoliberal capital: 'minds and *bodies* in a neoliberal groove'.
33 Stengers explains that slow science should 'enable scientists to accept what is messy not as a defect but as what we have to learn to live and think in and with. The symbiosis of fast science and industry has been privileging disembedded and disembedding knowledge and strategies, abstracted from the messy complications of this world. But messiness is returning with a vengeance. Ignoring it, dreaming of its eradication, we discover that we have messed up our world. I would then characterize slow science as the demanding operation which would reclaim the art of dealing with, and learning from, what scientists too often consider messy, that is, what escapes general, so-called objective, categories' ('A plea for slow science', 10).
34 Rabinow and Bennett, *Designing Human Practices*.
35 Compare Stengers, *Thinking With Whitehead*, 402–3, to Slavoj Žižek, *The Sublime Object of Ideology*, 131–2.
36 It should be noted here that, according to Whitehead, many of the problems in the history of philosophy – including Hume's causality and Kant's phenomena and noumena distinction – result from misunderstanding these two modes of perception. Both Hume and Kant, Whitehead contends, treat sense-perception as primary and reduce causal efficacy to a mere mental act. 'For according to these accounts', Whitehead maintains, 'causal efficacy is nothing else than a way of thinking about sense-data' (S 40). It is no wonder, then, that causality was reduced to a category of pure reason (Kant), on the one hand, and a habit of the mind (Hume), on the other. Whitehead, for his part, reverses this privilege and argues that causal feelings are primary and sense-perceptions are secondary. See *Symbolism: Its Meaning and Effect*, ch. 2, and *Process and Reality*, chs 5, 6 and 8. Even Hume, according Whitehead, could not maintain the 'purity' of sense-perception and had to admit the necessity of perceiving according to the mode causal efficacy. Quoting Hume, Whitehead explains, '"[I]f it be perceived by the eyes, it must be a colour; if by the ears, a sound; if by the palate, a taste; and so of the other senses." Thus in asserting the lack of perception of causality, he implicitly presupposes it. For what is the meaning of "*by*" in "*by* the eyes", "*by* the ears", "*by* the palate"? His argument presupposes that sense-data, functioning in presentational immediacy, are "given" by reason of "eyes", "ears", "palates" functioning in causal efficacy. Otherwise his argument is involved in a vicious regress' (S 51).
37 Stengers, *Thinking With Whitehead*, 404.

38 See Stengers' discussion of *felt theory* in *Thinking With Whitehead*, 405–22.
39 Stengers makes the important point that propositions are fully general in *Process and Reality*. In other words, propositions are not 'materials for judgment' that are meant to conform to the actual world (see PR 178). They are among the eight categories of existence: '(vi) Propositions, *or* Matters of Fact in Potential Determination, *or* Impure Potentials for the Specific Determination of Fact, *or* Theories' (PR 22).
40 For an illuminating account of Whitehead's metaphysical theory of the proposition in relation to the two-value logic that he maintains in the *Principia*, see Lucas, *The Rehabilitation of Whitehead*, 141–2.
41 See Pignarre and Stengers, *Capitalist Sorcery*, and Stengers, 'Reclaiming animism'.
42 For Whitehead, metaphysical abstraction always requires 'empirical' verification, which is to say, all elements of experience must be accounted for within the system. Whitehead tells us that, '[w]hatever is found in "practice" must lie within the scope of metaphysical description. When the description fails to include the "practice" the metaphysics is inadequate and requires revision' (PR, 13). Elsewhere, he explains that metaphysics 'starts from the ground of a particular observation . . . makes a flight into the thin air of imaginative generalization . . . and lands renewed for observation rendered acute by rational interpretation' (PR, 5). Also see Stengers, *Thinking With Whitehead*, 233–53.
43 See for example Fisher, *Capitalist Realism*, and Berardi, *After the Future*.
44 See Shaviro, *Without Criteria*, ch. 6; and Halewood, *Rethinking the Social Through Durkheim, Marx, Weber and Whitehead*.

Bibliography

Band, Yehuda B., and Yshai Avishai, *Quantum Mechanics with Applications to Nanotechnology and Information Science* (Burlington: Elsevier Science, 2013).

Beckman, Frida, 'The idiocy of the event: between Antonin Artaud, Kathy Acker and Gilles Deleuze', *Deleuze Studies*, 3:1 (2009), 54–72.

Bell, Jeffrey A., 'Scientism and the modern world', in Nicholas Gaskill and A. J. Nocek (eds), *The Lure of Whitehead* (Minneapolis: University of Minnesota Press, 2014), pp. 65–91.

Berardi, Franco, *After the Future*, ed. Gary Genosko and Nicholas Thoburn (Edinburgh: AK Press, 2011).

Brown, Wendy, 'Neo-liberalism and the end of liberal democracy', *Theory and Event*, 7:1 (2003), 15–18.

Cooper, Melinda, *Life as Surplus: Biotechnology and Capitalism in the Neoliberal Era* (Seattle: University of Washington Press, 2008).

Daston, Lorraine, and Peter Galison, *Objectivity* (New York: Zone Books, 2007).
Debaise, Didier, 'The living and its environments: Stengers reading Whitehead', *Process Studies*, 37:2 (2009), 127–39.
Deleuze, Gilles, *Difference and Repetition*, trans. Paul Patton (London: Continuum, 2001).
Deleuze, Gilles, and Félix Guattari, *What Is Philosophy?*, trans. Hugh Tomlinson and Graham Burchell (New York: Columbia University Press, 1994).
Endy, Drew, 'Nature is designed', in Alexandra Daisy Ginsberg, Jane Calvert, Pablo Schyfter, Alistair Elfick and Drew Endy (eds), *Synthetic Aesthetics: Investigating Synthetic Biology's Designs on Nature* (Cambridge, MA: MIT Press, 2014), pp. 7–86.
Esposito, Roberto, *Bíos: Biopolitics and Philosophy*, trans. Timothy Campbell (Minneapolis: University of Minnesota Press, 2008).
Fisher, Mark, *Capitalist Realism: Is There No Alternative?* (Winchester: Zero Books, 2009).
Flaxman, Gregory, *Gilles Deleuze and the Fabulation of Philosophy* (Minneapolis: University of Minnesota Press, 2012).
Foucault, Michel, *The Birth of Biopolitics: Lectures at the Collège De France, 1978–79*, trans. Michel Senellart (Basingstoke: Palgrave Macmillan, 2008).
Gaskill, Nicholas and A. J. Nocek, 'Introduction: an adventure of thought', in Nicholas Gaskill and A.J. Nocek (eds), *The Lure of Whitehead* (Minneapolis: University of Minnesota Press, 2014).
Ginsberg, Alexandra Daisy, Jane Calvert, Pablo Schyfter, Alistair Elfick and Drew Endy (eds), *Synthetic Aesthetics: Investigating Synthetic Biology's Designs on Nature* (Cambridge, MA: MIT Press, 2014).
Halewood, Michael, *Rethinking the Social Through Durkheim, Marx, Weber and Whitehead* (New York: Anthem Press, 2014).
Hardt, Michael, 'Affective labor', *Boundary 2*, 26:2 (summer 1999), 89–100.
Hardt, Michael, and Antonio Negri, *Empire* (Cambridge, MA: Harvard University Press, 2000).
Lambert, Gregg, *In Search of a New Image of Thought: Gilles Deleuze and Philosophical Expressionism* (Minneapolis: University of Minnesota Press, 2012).
Lazzarato, Maurizio, 'Immaterial labor', in Paolo Virno and Michael Hardt (eds), *Radical Thought in Italy: A Potential Politics* (Minneapolis: University of Minnesota Press, 1996).
Lazzarato, Maurizio, *The Making of the Indebted Man: An Essay on the Neoliberal Condition*, trans. Joshua D. Jordan (Los Angeles: Semiotext(e), 2012).
Lucas, George R., *The Rehabilitation of Whitehead: An Analytic and Historical Assessment of Process Philosophy* (Albany: State University of New York Press, 1989).

Marx, Karl, *Capital*, trans. David Fernbach (London: Penguin Classics, 1993).
McGill, Gael, 'Molecular movies . . . coming to a lecture near you', *Cell*, 133:7 (2008), 1127–32.
Mitchell, Robert, *Bioart and the Vitality of Media* (Seattle: University of Washington Press, 2010).
Pellizzoni, Luigi, and Marja Ylönen, *Neoliberalism and Technoscience: Critical Assessments* (Farnham: Ashgate Publishing, 2012).
Pignarre, Philippe, and Isabelle Stengers, *Capitalist Sorcery: Breaking the Spell*, trans. Andrew Goffey (Houndmills: Palgrave Macmillan, 2011).
Rabinow, Paul, and Gaymon Bennett. *Designing Human Practices: An Experiment with Synthetic Biology* (Chicago: University of Chicago Press, 2012).
Rajan, Kaushik Sunder, *Biocapital: The Constitution of Postgenomic Life* (Durham: Duke University Press, 2006).
Rajchman, John, *The Deleuze Connections* (Cambridge, MA: MIT Press, 2000).
Shaviro, Steven, *Post Cinematic Affect* (Winchester: Zero Books, 2010).
Shaviro, Steven, *Without Criteria: Kant, Whitehead, Deleuze, and Aesthetics* (Cambridge, MA: MIT Press, 2009).
Stengers, Isabelle, '"Another science is possible!" A plea for slow science', Inaugural Willy Calewaert Leerstoel lecture, Vrije Universiteit Brussel, 2012.
Stengers, Isabelle, *Cosmopolitics I*, trans. Robert Bononno (Minneapolis: University of Minnesota Press, 2010).
Stengers, Isabelle, *Cosmopolitics II*, trans. Robert Bononno (Minneapolis: University of Minnesota Press, 2011).
Stengers, Isabelle, 'Introductory notes on an ecology of practices', *Cultural Studies Review*, 11:1 (2005), 183–96.
Stengers, Isabelle, 'Reclaiming animism', *E-flux*, 36 (July 2012), available at http://www.e-flux.com/journal/36/61245/reclaiming-animism (last accessed March 2017).
Stengers, Isabelle, 'The cosmopolitical proposal', in Bruno Latour and Peter Weibel (eds), *Making Things Public: Atmospheres of Democracy* (Karlsruhe: MIT Press, 2005), pp. 994–1003.
Stengers, Isabelle, *Thinking with Whitehead: A Free and Wild Creation of Concepts*, trans. Michael Chase (Cambridge, MA: Harvard University Press, 2011).
Virno, Paolo, 'General intellect', *Historical Materialism*, 15:3 (2007), pp. 3–8.
Whitehead, Alfred North, *Modes of Thought* (New York: The Free Press, [1938] 1968).
Whitehead, Alfred North, *Process and Reality: An Essay in Cosmology*, corrected edition, ed. David Ray Griffin and Donald W. Sherburne (New York: The Free Press, [1929] 1978).

Whitehead, Alfred North, *Science and the Modern World* (New York: The Free Press, [1925] 1967).
Whitehead, Alfred North, *Symbolism: Its Meaning and Effect* (New York: Fordham University Press, [1927] 1985).
Whitehead, Alfred North, *The Concept of Nature* (Cambridge: Cambridge University Press, [1920] 1964).
Williams, James, 'Whitehead's curse?', in Nicholas Gaskill and A. J. Nocek (eds), *The Lure of Whitehead* (Minneapolis: University of Minnesota Press, 2014), pp. 249–66.
Wolfe, Cary, *Before the Law: Humans and Other Animals in a Biopolitical Frame* (Chicago: University of Chicago Press, 2013).
Žižek, Slavoj, *The Sublime Object of Ideology* (London: Verso, 1989).

11

Of Symbolism: Climate Concreteness, Causal Efficacy and the Whiteheadian Cosmopolis

CATHERINE KELLER

I

I was beginning to wonder how to cajole this tightly compressed little work – like a stubborn bud or a maddeningly well sealed package – to open up for me. *Symbolism*, no matter how long I have inhabited however much of Whitehead's thought, was not springing open, not suggesting much that I was not pressing upon it. I was again wondering if it did not, more than the major works of Whitehead, need to be abandoned to those more scholastically and specifically lured to it. And so I let myself admit my anachronistic wish for this serene booklet from 1927: that it yield a clue as to how symbols might better stir attention to climate catastrophe, to this 'hyperobject' that is global warming.[1] Whitehead had the wreckage of war and revolution in mind, not of the habitable earth. Still, I was hoping for some wisp of prophecy camouflaged by his Victorian charm.

What I do not need just now is to catch myself committing a warmly Whiteheadian version of the fallacy of misplaced concreteness, whereby his abstract commitment to the concrete gives me yet another pretext for deferring what he calls 'the sense of common purpose' (*Symbolism*), for displacing 'the instinct for action' (*Science and the Modern World*). How tempting it is for process thinkers to think we are *doing* something, doing the 'concrete', that we are good pragmatists deploying an 'activist philosophy'.[2] After all, we diagnose this fallacy in other systems – systems philosophical, theological, political, sexual, economic. But if the fallacy of misplaced concreteness counts as original sin for process thought, no wonder it is a constant temptation for process thinkers. The creepiness of sin in Augustine is, after all, its derivation from the good. Just to get you creeping in this labyrinth with me right now may just dig us deeper into an evasion.

While circling in this pre-writing solipsism, I bumped into a recent study of people's perceptions of changing weather. And suddenly the relevance of *Symbolism* flared into immediacy. But what came to light is neither theoretically straightforward nor practically reassuring. Yet the study's conclusion seemed to cry out for a Whiteheadian interpretation. And now I could focus my question for our conversation: does Whitehead help us rethink strategies for public education about global warming? The study may even illustrate the three modes of perception that preoccupy *Symbolism*. Nonetheless, the shadow question of practice falls upon the text: is there time? The text asks back: what does time *symbolise* – having time, being on time – within our rapidly altering planetary space? Apocalypse after all? Or the temporality, signalled by a recently Whiteheadian work of political philosophy, of a possible planetary politics?

II

Perhaps it will be most helpful for our discussion just to convey the matter of the study. It was written by three sociologists concerned with people's perceptions of the winter of 2012, which was anomalously warm (the fourth warmest on record for the contiguous US).[3] It addresses a situation in which partisan disagreement about the relationship between climate change and weather extremes has become routine. Comparing Gallup polling results from early March 2012 (just after the winter ended) with actual temperature data from the lower 48 US states, the researchers analysed people's perceptions of the warmth of the winter they had just lived through in light of the temperature anomalies that actually occurred. The essay suggests that the climate issue may have become so politicised that our very perceptions of the weather itself – not first of all our interpretation, but our sense and feeling of the actual weather – are subtly slanted by political identities and cues.

Their first result was not surprising: in general, people accurately perceived that their weather had been out of whack. In places where the winter was unusually warm, they said as much. 'The greater the deviation of winter 2012 temperatures from the 30-year winter temperature average in respondents' states, the more likely that respondents report local winter temperatures to be warmer than usual', notes the paper.

Of course temperatures predicted people's perceptions of temperatures. But then things got a bit weird. What surprised the researchers were the *other* factors that also shaped people's assessment of how warm it was. The researchers found that political party affiliation had an effect: 'Democrats [were] more likely than Republicans to perceive local winter temperatures as warmer than usual', the paper reports. And beliefs about global warming also predicted temperature perceptions. Subjects who were more likely to think that scientists agree about global warming, or to think humans are causing it, were also more likely to report that the recent winter had been 'warmer than usual'.

This is indeed a startling result. While many studies show the effect of deeply held beliefs on whether folk are open to climate science, you would not think that 'your politics and climate beliefs . . . change your experience of weather itself'.[4] Yet these data suggest to the researchers that whether people actually physically *feel* differently, or whether they remember and reconstruct their weather experiences differently, 'worldview' is playing a role. 'It suggests to me', explains study co-author Riley Dunlap, 'that people have begun to filter their fundamental perceptions of what is going on at least partly through a partisan frame'.[5] So folk are filtering the actual sensorium of temperature, at least as recounted in the form of recent memory, through ideology.

III

Remember when talk about the weather was the most innocuous kind of conversation, the most disarming, the way you could connect within a shared world with any stranger?[6] Now there is always the chance when one of us says 'Weird weather!' that the next line will be – 'It's gonna get weirder'. One mutters it. One doesn't want to get into the politics of climate change with every stranger. But the meta-weirdness of weather is what is of interest to this discussion: the alteration of sensual perception by political perspective. We might fervently hope for the converse: that folk's material experience of weather would be altering their perspectives.

In Whiteheadian terms, however, what does 'your experience of weather itself' signify? Such everyday experience never would have been simple, even if it were harmless. Any sense of the temperature of a winter day would be analysable as an instance of the 'three modes each contributing its share of components to our individual

Of Symbolism 211

rise into one concrete moment of human experience' (S 17). In other words, weather perceived in the mode of presentational immediacy would present as the bright surface, like a screen, of actual things extended in space, felt with the vividness of the sensation right now of warmth. *Pereunt*.[7] And weather perceived in the mode of causal efficacy would entangle the long-term environment of constant shiftings along with emerging patterns of a new change, all experienced in a bodily, murky, 'primitive' way. That efficacy conforms us to it all in the influx of the immediate past. An immediacy unlike the variegations of presentational immediacy, it is the pressure of the actual world acting upon us, reminding us we do not preexist it; a past felt affectively, 'vague, haunting, unmanageable' (S 43). *Et imputantor*. Laid to account. So here the experiences of the weather over time would bear traces of our own individual and collective inseparability from its evolutionary vagaries and our adaptations throughout the Holocene. Whitehead insistently repeats the metaphor of 'haunting' to capture this material–affective relationality. One cannot prevent Derrida's 'hauntology', with its political reconsideration of messianic time, from spooking the discussion, gesturing to the uninvited traces of past significations. (The spirit of Derrida is always welcome, if not here thematised.)

Surely then it is this 'world disclosed in its causal efficacy, where each event infects the ages to come, for good or for evil', that holds us to account for the evil of man-made [sic] global warming (S 47). But no such accounting is happening consciously or actively except through the third mode of symbolic reference. That 'world disclosed' lends at once pathos and passion, *pascho*, to the action of symbolic reference. And symbolism in Whitehead *is* precisely an activity – as is the world it discloses in the insistently not-Kantian 'direct experience' of causal efficacy: 'the world as an interplay of functional activity' (S 28). The world is not just the place of all practices: it is one great interactive practice. Activity, deed, *pragma*: here lies the deep pragmatism of the vision, the source of the activist intensities of process thought, especially as funnelled globally through Claremont.

Perhaps many have presumed something like the thought that haunts me: I remember John Cobb saying to me on a walk in the early 1980s: 'We need enough ecological crisis to raise public consciousness, but not so much that it is too late'. I realise that I have been hoping that one more Hurricane Katrina or Sandy, one more year of California drought, might trigger a national butterfly effect of responsible change. (A monarch butterfly, not yet extinct, flutters through the mind.) Certainly the butterfly effects of climate catastrophe, in its relentless feedback loops of causality, are well underway. But a corresponding transformation has not happened, is not happening. 'Symbolically conditioned action is action which is thus conditioned by the analysis of the perceptive mode of causal efficacy effected by symbolic transference from the perceptive mode of presentational immediacy.' This analysis, Whitehead insists, may be right or wrong, depending upon whether 'it does, or does not, conform to the actual distribution of the efficacious bodies' (S 80). Insofar as that 'analysis', with its fluttering degrees of consciousness, 'is sufficiently correct under normal circumstances, it enables an organism to conform its action to long-ranged analysis of the particular circumstances of its environment' (S 80).

One must, of course, amplify Whitehead's account beyond the organism to the vast human collective, but such a move is warranted by his own generalisations about the three revolutions (not to be correlated with the three modes). Then we may infer that the needed conformation of our shared material life to the 'actual distribution of the efficacious bodies' of the planetary collective carrying the effects of 'global warming' – the glaciers, forests, methane spumes, drought-ridden land, acidified oceans – is not happening. It would seem that our collective 'analysis' has gone wrong indeed, failing to fit, to conform to, the 'actual distribution'. The 'long-ranged analysis' has been available for decades, and has been confirmed by 97% scientific consensus in this decade. But despite the data and despite considerable if fitful public attention, our collective action is instead conforming with overwhelming fidelity to the dictates of neoliberal capitalism and its breathtakingly short-term analyses.

What I found disturbing in that recent study of weather perception is that it leaves no reason to expect that more ecological crisis close to home might make the difference – shift the swing vote, convert the critical mass, bring on the ecodemocratic paradigm waiting in the (left) wings. Nor can I reassure myself that, well, of course, there is nothing more prone to the fallacy of misplaced concreteness than our perception of the weather: whether in the illusion of 'pure succession' (Hume) or of 'simple occurrence' (Kant). Alongside whatever is perceivable about the weather is the communication of deliberate fallacies (I do not mean the literary 'intentional fallacy'; I mean malignant intentions) through such vivid screen-immediacies of presentation as Fox News. One may argue that, yes, naturally such causal efficacy isn't ever just about the climate. It is massively, overwhelmingly operative through the forces of communal practices, of religious imaginaries, of family relations and therefore of political ideology: in short, of 'reflex' actions that can only with great difficulty get interrupted and reinterpreted by alternative symbolic transfers from the spatial sensorium to the haunting past. But such analysis makes it all the more disturbing that 'the climate issue may have become so politicized that our very perceptions of the weather itself are subtly slanted by political identities and cues'.[8]

We already know that the pronouncements of frightening facts do not convince the folk who need convincing. But now we can no longer imagine that our traumatised ecology will be an efficacious ally in fighting climate denial. We should not hope that further bodily, affective, local experiences of climate crisis combined with the immediate presentations of global news will help. The urgency, the militancy, that the fight requires will be so intensely charged politically and so obstructed economically that the data may just continue to have the opposite effect of what we expect from it. So then – more empathic, contextually canny efforts to communicate with moderate, swingable constituencies? Isn't that what President Obama tried? Or – detach ecological science from politics and focus on longer-term education, while undermining the Christian Science denial that religiously intensifies climate denial and indifference to the sixth extinction? While the carbon-crazed global economy marches on, ignoring in flagrante its biocidal effects?

We don't have time.

IV

> The symbolic expression of instinctive forces drags them out into the open: it differentiates them and delineates them. There is then opportunity for reason to effect, with comparative speed, what otherwise must be left to the slow operation of the centuries amid ruin and reconstruction. (S 69)

Okay, breathe. We might gain the 'comparative speed' we need. 'Reason' may here signify collective sanity exercising symbolic reference for the common weal. Hence Cobb speaks now of the 'insanity' of the economic rationales, deemed purely rational, of our anti-ecological civilisation.[9] Such disclosive symbolic action would be the alternative to going down our present path, that of masking our ruination of the habitable planet with some greenwashing here, some renewable technology there, and hoping that we'll muddle through.[10] To get up to speed means dramatic transformation within what 'reason' (at least the scientific kind) tells us is pretty much one generation. As the last report from the Intergovernmental Panel on Climate Change (IPCC) (at this writing) makes unhysterically clear, we may have thirty years before breaching the warming threshold for irreversible climate catastrophe. What we collectively decide during these decades will make the difference, in Adrian Parr's more vivid rhetoric, between 'transformation and extinction'.[11]

It may not be helpful in contemporary conversation to diagnose the problem in terms of 'instinctive forces', redolent of a mechanistic model Whitehead has helped to undermine. But he is actually bringing emotion into play in a new way, anticipatory of current affect theory. The 'common emotions' that language elicits bind people together – unconsciously and 'instinctively', in conservative unquestionabilities – while at the same time serving as the instrument of freedom, critical thinking, the expression of difference. So the point is not to mobilise an affect-transcendent reason against instincts emotive and bodily. His main thesis, he tells us, 'is that a social system is kept together by the blind force of instinctive actions, and of instinctive emotions clustered around habits and prejudices' (S 68).

Is this not quite precisely capturing the way the ideological obstruction to the reception of climate science operates upon physical perception? Or perhaps the thesis works more precisely when read through his notion of reflex action, as 'a relapse towards a more complex type of instinct'. Neither instinct nor relapse is in itself 'wrong'. But they may be 'unfortunate'. They are unconscious reactivities. It comes down to 'a false symbolic analysis of causal efficacy', which in 'the absence of conscious attention' may build up formidable habits, practices, institutions, indeed whole social orders. But democrats, and women, for instance, also perceive weather through a political lens that is therefore already a reflex. Are we any less prone to false analysis? One can just note that the symbolism ground into one lens of perception represses attention to the planetary ecology while the other presses for it (S 81).

Mindfulness may be one currently useable paraphrase of 'reason', if we wish to signify a thinking that critically exposes habits destructive to human flourishing, to the flourishing of all humans, and that in its course of thinking is learning to attend intellectually and practically to the entanglement of humans in non-human systems. Whitehead more than any philosopher helps us to materialise mindfully the *universe* of ethical universalism. The reasonably anthropocentric Kantian cosmopolitanism here morphs, through an epochally updated science, into the fully cosmological cosmopolis.

Whitehead notes that though 'reason too often fails', this is no excuse 'for the hysterical conclusion that it never succeeds'. He likens it to the weak force of gravity, 'in the end the creator of suns and of stellar systems'. Of course, this reason is haunted by the symbol, avoided in these Virginia lectures, of the God whose 'primordial nature' exercises the gentle and frequently ineffectual power – so very reasonable – of persuasion. But he links reason in this text less to slow democratic process than to 'the disruptive tendency due to novelties', and so to the three revolutions that made democracy possible. (Though hardly yet actual, comes the hauntological whisper.) This cosmically slow and patient force appears at the same time as the source of radical novelty in human self-organisation. So 'symbolically conditioned action' really does get into action (S 69–78).

The final image of *Symbolism* invites, not romantically but uncringingly, in all mindfulness of the risk, revolutionary action. It leads us 'to recognize that the major advances in civilization are processes which all but wreck the societies in which they occur: like unto an arrow in the hand of a child' (S 88). Contemporary, politically progressive readers can hardly miss the radical gesture. It is not out of synch with Whitehead's corpus. It echoes in Whitehead's meditations on novelty – indeed, on revolt and the rejection of 'mere unrelieved preservation of the past' – in the chapter on 'Ideal opposites' in *Process and Reality*, in tensive contrast with 'terror at the loss of the past'.

We do urgently need a major civilisational advance. Only the irruption of novel modes of self-organisation for our species in its dense interplay with its living planet will make possible an 'ecological civilization'.[12] This is to say that a viable planetary cosmopolitics will wreck the civilisation – and much of the undeniable good it yields – of the neoliberal capitalism that has since 1980 taken control of the economic globe. And such a cosmopolis cannot come about now – not on time – through democratic meandering. If it does happen it will happen only by way of the impatient fires of the new – a revolt whose way process thought is labouring to prepare, but that can be pulled off only by a generation furious at what we have done to its future.

Yet I fear that the symbolic reference Whitehead intends – in the discourse of juvenescent spontaneity and of revolutionary devastation – may itself ring 'wrong' for the present context. Not because it welcomes a certain violence. But such language of the necessary wreckage of self-conserving orders may misfire in the context of climate change. In this time of the Sixth Extinction, the unconscious recklessness of the child seems better to symbolise those advances we must now somehow undo: the progress of carbon-driven industrial civilisation, with its apocalyptic scale of wreckage. Much of the damage to the planet is collateral, inadvertent, produced by feedback loops, die-offs, spumes, shifts that were never intended: like unto the arrow . . .

Nonetheless, we need the speed of that arrow and so after all – if the change is to be more and other than academic – the raw risk it poses. Of course, neither the three revolutions he mentions nor the Marxist ones he does not would have been possible without intensive intellectual activity. Yet we may wonder if there is *time* for the symbolic transferences of a fresh, marvellously earth-calibrated reason to take *place*. And yet without a collective resymbolisation of our common life, won't the ongoing denialism, circling between capitalist rationalisation and reactionary Christianity, clench the deal? Do we not then face some final aeon of truly thoughtless violence, trumping the possibility of ecological civilisation and so of civilisation, scapegoating already victimised populations and warring for the last resources?

V

Still practising the archery of *Symbolism,* however, let me solicit the hope and the help of a political philosophy that rethinks time itself in acute mindfulness of climate catastrophe. It does so with a cosmically imbricated militancy and a freshly Whiteheadian voice. William Connolly, long a deft synthesist of Deleuzian and Jamesian thought, has for just as long as a non-theist practised a 'presumptive generosity' that enables conversation with non-resentful forms of theism. His pragmatic commitment to building alliances wide, diverse and plastic enough to counter the 'evangelical-capitalist resonance machine' has combined with his interest in biological models of complex self-organisation to most recently launch him upon a serious study of Whitehead, whose theism he neither ignores nor amplifies. But above all it is his realisation of the dauntingly dynamic interaction of neoliberal economics with climate change that has spurred this new engagement, and its particular thinking of temporality.

Connolly offers in *The Fragility of Things* a Whiteheadian rendition of what he calls 'temporal forcefields' (FT 8). The key chapter

Of Symbolism 217

for our purposes is called 'Process philosophy and planetary politics'. He opens with an allusion to the Weavers singing in the 1950s 'The future's not what it used to be'. He adds, 'What's more, it never was'. That is not as reassuring a generalisation as it seems. It means to him that

> dangers to the human estate itself press on the horizon during an era when capitalism has intensified and when encounters between it and a variety of nonhuman force fields with independent powers of metamorphosis have once again become dicey. It also means that to understand those dangers and possibilities we may need to recraft the long debate between secular, linear and deterministic images of the world on the one hand and divinely touched, voluntarist, providential and/or punitive images on the other. Doing so to come to terms more closely with a world composed of interacting force fields set on different scales of chronotime composing an evolving universe open to an uncertain degree. (FT 149)

He notes that such a world, with its peculiarly endangered future, returns us to an ancient sensibility, that is, to disturbing levels of vulnerability to the forces of non-human nature, personified in Hesiod as divine. I might mumble that Hesiod's antagonistic heroism and misogynist account of Gaia hardly oriented civilisation to an earthwise future. But Connolly's larger point is crucial: 'the planetary fragility of things is increasingly sensed, as many protest against acknowledgement of that very sense to remain loyal to traditions of belonging woven into their bodies, role performances, and institutions' (FT 172). His sense of the embodied performativity of these traditions, and the strength of the affective commitment to them, is thought that is in resonance with Whitehead's sense of causal efficacy. It thus implies also the analysis of symbolic reflex action. This sheds a different light (beyond the normal analysis of their reactionary dominion dogma) on the widespread denial of climate change among the mass of conservative Christians. These are not the 1%, but those whose dominion tradition lends them dignity and strength amid the travails of their lives. And one senses the protest too among working-class individualists proud of dominating and transcending 'nature', when so much else eludes their mastery. These folk have for generations had their symbolic activity 'transferred' by the canny politico-economic doctrines of the right.

This protest against fragility, against mortality itself, comes home to roost, however, in the new and massive hyperobject. Climate change reminds us of how we are all wired – instinctually – to fights and flights of an immediate sort. Without disciplined analysis we are not readily able to respond to the strained sort of abstraction that climate science, with its 350 ppm, its complex currents where warming causes extreme cold, its century-long – or even decade-long – time frame demands of us. Certainly, popular loyalty to prior patterns, with their performative, prehensive embodiments, suggests why environmentalist litanies of the fearful facts remain futile. So, as Connolly points out, we often get instead 'bellicose political movements of denial and deferral . . . Witness the media attacks on scientists of climate change and proponents of sustainable energy' (FT 173). He is investigating how economic neoliberalism is successfully manipulating our civilisation through the symbolism of a subliminal theodicy: global capitalism lulls its constituencies to trust in the ultimate rationality of the market. Sure, there are losses, challenges and sacrifices – but have faith in the promise: the market remains the best of all possible worlds, in which true believers will be richly rewarded. The deserving win.

In Connolly's analysis of the political economy and its effects across multiple non-human force fields, it is ye olde fallacy of misplaced concreteness, now read in conjunction with quantum entanglement theory, that offers him the needed clue to our condition. 'For Whitehead', he writes:

> misplaced concreteness means more broadly the tendency to overlook entanglements between energized, real entities that exceed any atomistic reduction of them, as when a climate pattern and ocean current system intersect and enter into a new spiral of mutual amplification, or when a cultural disposition to spiritual life befuddles the academic separation between an economic system and religion by flowing into the very fiber of work motivation, consumption profiles, investment priorities, and electoral politics. (FT 154).

One hears repeatedly the energetic pulsion of Connolly's lists of concrete examples (the opposite of abstract listlessness).[13] Note that in *Symbolism* it was specifically the fallacy of misplacing the concreteness of *time*, rendering it as pure succession and as simple location, that got Whitehead's goat. The delusions of disentangled selves pursuing their independent projects along a straight line to their comeuppance may be what has so dangerously misplaced all of our futures. But then I must also ask whether the undeniably apocalyptic rhetoric of 'too late', of running out of time, a rhetoric I sometimes catch myself practising, might not get trapped in its own abstraction – where time in the space of the earth gets unified by global warming into a single overdetermined slide towards doom.

VI

I'll watch out for that. Having also been riveted to the quantum metaphor of entanglement, having traced an 'apophatic entanglement' shadowing and complicating human and non-human interactivities at every register, I will not now surrender to the simplification of an eco-apocalypse. Not even for the sake of a wider, more effectual discourse. Its prophetic warning will backfire as soon as it becomes certain of itself. The value of a symbol over a mere fact is nothing without its margin of indefiniteness – the alternative to the false certainty of misplaced concreteness. The mysteriousness of that margin amplifies in proportion to its existential importance.

In its own attention to complexity – not lacking in allusion to mystery – Connolly's interweaving of Whitehead with physical entanglement works to energise a planetarily responsible pluralism.

> Misplaced concreteness thus downplays both entanglements and processes of self-organization on the way, depreciating how every thing is enmeshed with others and metamorphizes according to the time scale appropriate to it. Such an image of multiple entanglements does not, therefore, devolve into a kind of organic holism, for that move would subtract the element of real creativity from the universe. (FT 154)

In honour of James, he calls his Whiteheadian move 'protean connectionism' (FT 154). He draws then from Whitehead a spirituality that (with a Nietzsche supplement) pushes free of *ressentiment*, particularly religious resentment of the aleatory risk and unavoidable mortality of life.

It is Whitehead's symbol for ultimacy – 'real creativity' – that Connolly wants now to mobilise 'whenever a reductionist explanation of change fails, when something new is added to a preexisting environment, when the newness involves a degree of self-organization on the part of at least one of the entities involved' (FT 158). It applies whenever a new equilibrium is promoted through self-organisation by way of what Connolly has called, with Terrence Deacon, 'teleodynamic searches' (FT 85). This teleodynamism names an alternative to any nihilist purposelessness as well as to the misplaced concreteness of eschato-teleological timelines.

If Connolly finds sustenance in the category of creativity, it is particularly in its aesthetic element. 'The creative relation, to Whitehead, operates by attraction and repulsion within and between interacting entities; otherwise there would be little power of an entity to maintain itself.' So we find here the prehensive dynamism of causal efficacy driving, haunting, the symbol activities that surface the vitality of the world.

> It also means . . . that an aesthetic element is in play within relations of ingression, prehension, and concrescence. This aesthetic element is not merely operative in human relations or in the human relations to things; it is involved in several thing-thing relations too. (FT 158)

He lifts up the prehensive feeling that links his work so closely to Jane Bennett's and to the intersections of the new materialism with affect theory in general. And he picks up on the possible proximity in Whitehead of the *beautiful* to the *fragile* (FT 159). Perhaps it is no coincidence that as the fragility of things is coming into focus, so is 'the universe of things'.[14] Connolly might agree that 'the world is indeed, at its base, aesthetic. And through aesthetics, we can act in the world and relate to other things in the world without reducing it and them to mere correlates of our own thought.'[15]

We may press then for a planetary politics that hails the beauty of the world as a source of refreshment – and so not of distraction and commodification, but of renewal of the militant struggle for sustainable and just practices always exceeding human relations. 'You engage the planetary dimension of politics when you explore how ocean current flows, climate patterns, regional patterns of drought and flood, . . . hurricane patterns and so on impinge upon us and how cultural processes impinge upon them' (FT 176). He finds therefore a

> cosmic dimension folded into contemporary politics, in part because it speaks to a time when several planetary force fields become entangled densely with several aspects of daily life, in part because our capacities to explore and respond politically to such imbrications with affirmative intelligence are severely challenged, in part because dangerous existential dispositions surge and flow again into defining institutions of late modern life. (FT 178)

We find here unfolding in a richly American naturalist/process symbolics, intensified with Nietzschean and Deleuzian Continentalism, what is fleetingly named 'the cosmopolitical dimension' (FT 171). He creates thereby an opening in political theory to the cosmos of cosmopolis. It is a mindful opening into the spatio-temporality of the Sixth Extinction and of global warming – and so at the same time to a new pluri-temporality of political alliance.

VII

An emergent symbolism of cosmopolitics means to energise – in time – down-to-earth work with local communities and piecemeal assemblages, involving translation across a wider-than-comfortable political spectrum. I began with a discouraging recent study. Let me mention another study, based in urban New York, which in the face of such data proposes 'framing for relevance', 'participation' and 'systems thinking' as strategies for climate change education and preparation.

> For more conservative audiences, environmentally friendly behaviour is much more attractive when it is framed as an issue of economic or energy security, such as seeking independence from foreign oil, or an act of patriotism, such as buying goods made in the United States. For audiences that experience oppression, framing climate change as an issue of justice can be a good way to tap into what people are already passionate about and personally affected by, because economically disadvantaged communities are often more heavily affected by extreme weather events.[16]

Such diverse framings get the haunting sense of shared vulnerability (*imputantur*) symbolically referenced across a significant margin of politically or religiously conservative communities that characterise much of the US middle. The glocality of pluritemporal education for action has nothing to do with ideological purity. This does not mean that, at the level of theory, the ecosocial justice that drives a planetary resistance to neoliberal capitalism becomes interchangeable with pragmatic local and localising rhetorics (such as patriotism).

In the meantime, the mix of liberal/progressive publics who are already to varying degrees concerned with the social impact of climate change have hardly yet joined forces – though for instance the People's Climate March of 2014, with its strong representation of people of colour, of labour unions, of religions, performed a great and celebratory symbolic action, greatly promising of the needed diversity.[17] Connolly ultimately suggests the need for a galvanising aim, not a final telos but a 'beacon'. Perhaps 'those who both appreciate the fragility of things and hear a call for democratic militancy at several sites can be further energized by connecting these critiques, proposals, and actions to such a beacon' (FT 194). We may recognise in the teleodynamism of this 'beacon' the symbolic activity that is intensified by mindful entanglement of both modes of perception. This call, which he recognises as a Whiteheadian 'lure', can 'infuse a sensual dimension' and so a presentationally immediate causal efficacy into the polytemporal public mix. Concretely he is proposing a general strike across several countries. The overriding goal of such a graded strike is:

> to press international organizations, states, corporations, banks, labor unions, churches, consumers, citizens, and universities to act in concerted ways to defeat neoliberalism, to curtail climate change, to reduce inequality, and to instill a vibrant pluralist spirituality into democratic machines that have lost too much of their vitality. (FT 195)

The wildly broad-spectrum multiplicity of Whitehead's cosmology finds here a promising political concreteness. That such political philosophy can with no theistic avowal call upon a range of spiritual practices, theistic and not, suggests the potential of its ecological ecumene for a planetary politics. Such a planetary ecumene (or shall we spell it ecomene?) will theologically approximate the rigorous pluralism of Roland Faber's 'divine manifold'.[18] Its radical multiplicity redistributes divine energies in theoplicity, energising in theory the difficult planetary connectivities waiting to happen in practice. If an international interlinkage of movements will welcome the haunting prophetic ancestors, the ancient messianic echoes, the peace-making wisdoms, inasmuch as they reciprocally welcome the secular radicalisations of democracy, great new forms of symbolic action become possible. When will they strike?

VIII

From the perspective of a theologically apophatic entanglement in the time of the space of global warming, must we reconsider the symbols of the old apocalypse? They are so reflexively entrenched and divergently alive across the troublesome width of the religious spectrum that one must now and again interpret rather than discard them.[19] Now I would focus on the role of the planet in the Book of Revelation, with all its hallucinogenic narrative of floods and fires and droughts and the death of the life of the sea symptomatic, if not of neoliberalism, of its economic ancestor, the 'whore of Babylon', depicted as wallowing in the wealth and labour of the world. 'Woe, woe, woe to the inhabitants of the earth' (Rev. 8: 13). And yet also: 'awake and strengthen what can still be saved' (Rev. 3: 2). Only through its *apophasis* can *apokalypsis* disclose rather than merely close possibility. But the text itself, vengeful and misogynist in its teleological determination, nonetheless does not announce 'the end of the world'. Just the cosmically scaled wreckage that precedes the cosmopolitan New Jerusalem. An urban park, to be sure, or 'megalopolis that is a continent-sized shopping mall',[20] but neither Heaven nor The End.

It seems evident that we will not avoid some significant fulfilment of its prophecy of doom. And the apocalyptic tradition will not have been innocent of the outcome. Whether some descendent of the eschatological hope of a just and convivial cosmopolis will be realised under conditions of mitigation and adaptation remains unknown. The apocalyptic symbols are dangerous; for that very reason they are not to be avoided. Whitehead was not thinking of such characters as the roaring Lamb or whore of Babylon when he wrote that 'each symbolic transference may involve an arbitrary imputation of unsuitable characters' (S 87). Perhaps, because biblical symbols have as much strength as any political symbols, they can help with some of the public on the swingable right. And that is a public pre-eminently concerned, even within tight communities of family and church, with the individual – in an individualism formed under neoliberal economic pressure intensified for several decades by anticommunism. Under apocalyptic pressure, it will be wise to support some version of this disarming ontology that never opposes the social or ecological collective to the intrinsically worthwhile individual:

> The world is a community of organisms; these organisms in the mass determine the environmental influence on any one of them; there can only be a persistent community of persistent organisms when the environmental influence in the shape of instinct is favorable to the survival of the individuals. (S 87)

That Whitehead's radical relationalism never washes out difference but intensifies it, that the singular subject happens – if only for a moment – may actually make his theory of symbolism useable and useful in shifting the individualism of the US public. The vision of the world as a community of organisms is no longer a matter of aesthetic preference or scientific debate but of urgent necessity – for the survival not of mere individuals, but of the life-systems in which they 'dividually' happen. In this world which is an interplay of functional activity, and as such a community of communities of communities, we find ourselves 'amid a democracy of fellow creatures' (PR 50). To be sure, our species has failed to evolve soon enough an ecosocial format for such a democracy. So the threatened space and shrinking time of our century will expose our participation in temporal force fields and spatial entanglements that until now have only haunted the spiritual margins. That exposition can serve as revelation of needed lures, beacons to our better instincts. Is there time to actualise their promise? Or will the 'unsuitable characters' sabotage our best efforts?

Any pretence of an answer will be its own symbolic action, its own self-fulfilling prophecy. So why quit symbolising 'the sense of common purpose' now, just as new relational radicalizations, affective materialisations, multi-racial-feminist-queer-eco-democratic-politics, are transgressing disciplinary and movement boundaries, making possible, just possible, some unpredictable concrescence, perhaps a coalescence, heavens, maybe even a coalition?

Notes

1 Morton, *Hyperobjects*.
2 Massumi, *Semblance and Event*.
3 McCright et al., 'Increasing influence of party identification'.
4 Mooney, 'Do Democrats and Republicans actually experience the weather differently?' Mooney cites McCright et al., 'The impacts of temperature anomalies and political orientation on perceived winter warming'. A quick sampling of recent research with the help of my research assistant Kyle Warren suggests this study reinforces others; and that there is little empirical evidence of the opposite.
5 Mooney, 'Do Democrats and Republicans actually experience the weather differently?'
6 Keller, 'Talk about the weather', 30–49.
7 '"Pereunt et imputantur" is the inscription on old sundials in "religious" houses: "The hours perish and are laid to account." Here "Pereunt" refers to the world disclosed in immediate presentation, gay with a thousand tints, passing, and intrinsically meaningless. "Imputantur" refers to the world disclosed in its causal efficacy, where each event infects the ages to come, for good or for evil' (S 47).
8 Mooney, 'Do Democrats and Republicans actually experience the weather differently?'
9 Cobb, Jr, *Spiritual Bankruptcy*.
10 Klein, *This Changes Everything*.
11 Parr, *The Wrath of Capital*, 3. Cf. Keller, *Cloud of the Impossible*, 282.

12 'Seizing an alternative: toward an ecological civilization', 4–7 June 2015, https://www.ctr4process.org/whitehead2015 (accessed).
13 Keller, 'Omnipotence and *The Fragility of Things*'.
14 Shaviro, *The Universe of Things*.
15 Shaviro, *The Universe of Things*, 156.
16 Snyder et al., 'City-wide collaborations for urban climate education', 106.
17 The People's Climate March of 2014 involved 2,646 events planned in 162 countries. The largest march, held in New York City, included 400,000 people (50,000 of whom were college students) and 1,574 organisations. Among the marchers representing diverse global and local communities were: Gibran Raya, the director of an indigenous dance troupe; Favianna Rodriguez, the director of Culture Strike, an immigration rights organisation; Uncle Angaangaq Angakkorsuaq, a shaman from Greenland; Margaret Lokawua, an indigenous leader from rural north-east Uganda; a contingent of Hurricane Sandy victims; and leaders of Avaaz, a global civic organisation. See Kormann, 'The three hundred thousand'.
18 Faber, *The Divine Manifold*.
19 I called such a move the counterapocalypse, rather than the anti-apocalypse. Keller, *Apocalypse Now and Then*.
20 Moore, *Untold Tales from the Book of Revelation*, 225.

Bibliography

Cobb, Jr, John B., *Spiritual Bankruptcy: A Prophetic Call to Action* (Nashville: Abingdon Press, 2010).
Connolly, William E., *The Fragility of Things: Self-Organizing Processes, Neoliberal Fantasies, and Democratic Activism* (Durham, NC: Duke University Press, 2013).
Faber, Roland, *The Divine Manifold* (New York: Lexington Books, 2014).
Keller, Catherine, *Apocalypse Now and Then: A Feminist Guide to the End of the World* (Boston: Beacon Press, 1996).
Keller, Catherine, *Cloud of the Impossible: Negative Theology and Planetary Entanglement* (New York: Columbia University Press, 2014).
Keller, Catherine, 'Omnipotence and *The Fragility of Things*: the cosmopolitics of William Connolly', *Theory and Event*, 18:3 (2015), n.p.
Keller, Catherine, 'Talk about the weather: the greening of eschatology', in Carol J. Adams (ed.), *Ecofeminism and the Sacred* (New York: Continuum, 1993), pp. 30–49.
Klein, Naomi, *This Changes Everything: Capitalism vs. The Climate* (New York: Simon and Schuster, 2014).
Kormann, Carolyn, 'The three hundred thousand', *The New Yorker*, 23 September 2014, available at http://www.newyorker.com/tech/elements/three-hundred-thousand-climate-march (last accessed March 2017).

Massumi, Brian, *Semblance and Event: Activist Philosophy and the Occurrent Arts* (Boston, MA: MIT Press, 2011).

McCright, Aaron M., Riley E. Dunlap and Chenyang Xiao, 'Increasing influence of party identification on perceived scientific agreement and support for government action on climate change in the United States, 2006–12', *Weather, Climate, and Society*, 6 (2014), 194–201.

McCright, Aaron M., Riley E. Dunlap and Chenyang Xiao, 'The impacts of temperature anomalies and political orientation on perceived winter warming', *Nature Climate Change*, 4 (2014), 1077–81. Available at http://www.nature.com/articles/nclimate2443.epdf (accessed 27 November 2014).

Mooney, Chris, 'Do Democrats and Republicans actually experience the weather differently?' *Washington Post*, 24 November 2014, available at https://www.washingtonpost.com/news/wonk/wp/2014/11/24/do-democrats-and-republicans-actually-experience-the-weather-differently (last accessed March 2017).

Moore, Stephen, *Untold Tales from the Book of Revelation: Sex and Gender, Empire and Ecology* (Atlanta: SBL Press, 2014).

Morton, Timothy, *Hyperobjects: Philosophy and Ecology After the End of the World* (Minneapolis: University of Minnesota Press, 2013).

Parr, Adrian, *The Wrath of Capital: Neoliberalism and Climate Change Politics* (New York: Columbia University Press, 2013).

Shaviro, Steven, *The Universe of Things: On Speculative Realism* (Minneapolis: University of Minnesota Press, 2014).

Snyder, Steven, Rita Mukherjee Hoffstadt, Lauren B. Allen, Kevin Crowley, Daniel A. Bader and Radley M. Horton, 'City-wide collaborations for urban climate education', in Diana Dalbotten, Gillian Roehrig and Patrick Hamilton (eds), *Future Earth: Advancing Civic Understanding of the Anthropocene* (Washington, DC: American Geophysical Union: 2014), pp. 103–9.

Whitehead, Alfred North, *Process and Reality: An Essay in Cosmology*, corrected edition, ed. David Ray Griffin and Donald W. Sherburne (New York: The Free Press, [1929] 1978).

Whitehead, Alfred North, *Science and the Modern World* (New York: Macmillan, 1925).

Whitehead, Alfred North, *Symbolism: Its Meaning and Effect* (New York: Capricorn Books, [1927] 1959).

Notes on Contributors

Jeffrey Bell is a Professor of Philosophy at Southeastern Louisiana University. His work focuses on contemporary Continental philosophy and its intersections with early modern philosophy, especially Hume and Spinoza. Bell has also written extensively on Whitehead.

Roland Faber is the Kilsby Family/John B. Cobb, Jr., Professor of Process Studies at Claremont School of Theology, Professor of Religion and Philosophy at Claremont Graduate University, Co-Director of the Center for Process Studies, and Executive Director of the Whitehead Research Project. His fields of research and publication include Whitehead's philosophy, process philosophy and process theology; (de)constructive theology; post-structuralism (Gilles Deleuze); transreligious discourse (epistemology of religious relativity and unity) and interreligious applications (e.g., Christianity, Buddhism, Bahá'í faith); comparative philosophy and mysticism (Meister Eckhart, Nicholas of Cusa, Ibn 'Arabi); and theopoetics (an approach to post-structuralist and process theology, which addresses the liberating necessity of multiplicity). His publications include *God as Poet of the World* (2008), *Event and Decision* (2010), *Beyond Metaphysics?* (2010), *Secrets of Becoming* (2011), *Butler on Whitehead* (2012), *Theopoetic Folds* (2013), *Beyond Superlatives* (2014), *The Allure of Things* (2014), *The Divine Manifold* (2014) and *Living Traditions and Universal Conviviality* (2016).

Michael Halewood is a senior lecturer at the University of Essex. His research interests lie at the intersection of social theory and philosophy. He is the author of two monographs: *A. N. Whitehead and Social Theory: Tracing a Culture of Thought* (2011) and *Rethinking the Social Through Durkheim, Weber, Marx and Whitehead* (2014). He has also written on Badiou, Butler, Dewey, Deleuze and Irigaray, as well as topics such as the tuning of musical instruments and modernity (*History of Human Sciences*), the form and value of things (*British Journal of Sociology*), entropy and death (*Social Science*) and conceptions of the self in those diagnosed with Alzheimer's disease (*Sociological Review*).

Luke Higgins has interests that lie at the intersection of constructive theology, process thought, Continental philosophy, science studies and ecological philosophy. He received his doctorate in theological and philosophical studies from Drew University, where he studied with constructive theologian Catherine Keller. His dissertation, entitled *The Time of Ecology: Theological Cosmology for a Postmodern Earth*, uses the philosophy of Whitehead, Deleuze and Bergson to think towards an approach to theological cosmology capable of affirming spontaneous, creaturely self-creativity, on one hand, and divinely inflected 'trans-temporal' trajectories of meaning and value, on the other hand. His constructive synthesis moves towards a panentheistic, ecotheological cosmic Christology deeply critical of 'macro-teleological' concepts of cosmic design. His current research aims at articulating an ecological approach to religious experience grounded in a speculative, experimental method adapted from Deleuze and Whitehead, among others. He currently serves as lecturer in philosophy at Armstrong University.

Catherine Keller is Professor of Constructive Theology at the Theological School of Drew University. In her teaching, lecturing and writing, she develops the relational potential of a theology of becoming. Her books reconfigure ancient symbols of divinity for the sake of a planetary conviviality – a life together, across vast webs of difference. Thriving in the interplay of ecological and gender politics, of process cosmology, post-structuralist philosophy and religious pluralism, her work is both deconstructive and constructive in strategy.

Sheri Kling earned her PhD and MAR in Religion from Claremont School of Theology and holds a master of theological studies from the Lutheran School of Theology at Chicago. Her research includes integrating process theology/philosophy with the spiritual practice of Jungian dream work to facilitate transformation. She is a member of the American Academy of Religion, an accomplished songwriter and recording artist, and recently accepted a position at the University of the South in Sewanee, TN.

Hyo-Dong Lee is Associate Professor of Comparative Theology at Drew University Theological School and its Graduate Department of Religion. A native of South Korea, he holds a PhD from Vanderbilt University and is the author of *Spirit, Qi, and the Multitude: A Comparative Theology for the Democracy of Creation* (2014).

Beatrice Marovich is a writer and academic who teaches in the Department of Theological Studies at Hanover College. She earned her PhD at Drew University's Graduate Division of Religion in 2014. Her research and writing are ecologically and speculatively oriented. Before returning to graduate school, she was a newspaper reporter at a hardscrabble daily on the coast of Maine. She has a BA in Spanish and comparative literature from the University of Michigan, Ann Arbor (2004). She is an alumna of the Salt Institute for Documentary Studies, where she studied non-fiction writing and editing (2005). She earned her MA at the Vancouver School of Theology, in British Columbia (2009), and entered the PhD programme in Philosophy and Theology at Drew University's Graduate Division of Religion in the autumn of 2009.

Adam Nocek is an assistant professor in the philosophy of technology and science and technology studies in the School of Arts, Media and Engineering at Arizona State University. He works at the intersections of Continental philosophy and science studies, digital culture and aesthetics, and design and technoscience, and has published numerous essays on media theory, artificial life, process philosophy (especially Whitehead and Deleuze), architecture, and the history of biotechnology. Nocek is the co-editor of the collection *The Lure of Whitehead* (2014) and of a special issue of the journal *Inflexions*, titled 'Animating Biophilosophy' (2014). He is currently working on a manuscript titled *Animating Capital: Molecules, Labor, and the Cultural Production of Science*. Nocek is also the founding director of the Laboratory for Critical Technics (LCT) at Arizona State University.

Joseph Petek is a doctoral student in process thought at Claremont School of Theology. He is the Chief Archivist of the Whitehead Research Project, Assistant Series Editor for the *Critical Edition of Whitehead* and English Coordinator for the *Balkan Journal of Philosophy*. His research interests include process thought, philosophy of death and personal identity, and virtual worlds.

Keith Robinson is engaged in teaching and research concerned primarily with three main areas. The first is the European traditions of thought that emerge from Kant and post-Kantian – especially nineteenth- and twentieth-century Continental thinkers (Nietzsche, Foucault, Deleuze). Secondly, he has strong interests in modern process philosophy (James, Bergson, Whitehead). Finally, he is interested in the interconnections between these two areas. This last area of research has revolved around a critical exchange between process philosophy and perspectives drawn from post-structuralist and phenomenological thinkers. It has centred principally on temporal themes, especially the concepts of 'event' and 'process', across a range of contexts and problems. He contributed a chapter entitled 'The Event and the Occasion: Deleuze, Whitehead and Creativity' in N. Gaskill and A. Nocek (eds), *The Lure of Whitehead* (2014) and his most recent book is the edited collection *Deleuze, Whitehead, Bergson: Rhizomatic Connections* (2009).

Steven Shaviro is the DeRoy Professor of English at Wayne State University. He is the author of *Connected, Or, What It Means to Live in the Network Society* (2003), *Without Criteria: Kant, Whitehead, Deleuze, and Aesthetics* (2009), *Post-Cinematic Affect* (2010) and *The Universe of Things: On Speculative Realism* (2014). He blogs at 'The Pinocchio Theory' (http://www.shaviro.com/Blog).

Index

abstraction, 15, 19, 22–3, 33, 35, 36, 48, 54n, 58–61, 67, 88, 176–7, 191–2, 198, 204n
actual entity, 2, 3, 73n, 96, 102–4, 111–12, 127–9, 131, 149, 152, 161–2, 166
actual occasion *see* actual entity
actuality, 3, 62, 103–5
Adventures of Ideas, 69, 74n, 125, 172, 183
aesthetics, 103–4, 220
analytic philosophy, 15
animal thought, 147–53, 156–9, 165; *see also* perception: non-human
apocalypse, 219, 222–3
archetype, 132–40
Aristotle, 147

Barthes, Roland, 85–7
Beckman, Frida, 202n
becoming, 31, 38, 43, 46–7, 50–2, 70, 111, 118
becoming imperceptible, 43–5
Benveniste, Émile, 158
Bergson, Henri, 183, 184n, 185n
bifurcation of nature, 17–18, 29–30, 32–3, 39, 45, 81, 87, 172, 175
and transcendence, 39
body, 14, 21, 32–33, 42, 44, 67, 111, 138
Buddhism, 66–7
Butler, Judith, 59–61, 63

Caesar, 57–8
Camp, Elisabeth, 148–9, 156–9
Cantor, Paul, 100
capitalism, 90, 92–4, 170–1, 179, 189–90, 193–4, 198–9, 200n, 212, 216, 218, 221, 223
Cassirer, Ernst, 4
causal efficacy, 2, 15–26, 30, 34–6, 62–3, 66, 110, 120n, 130–1, 138–9, 152–3,

160–1, 173–5, 177, 179, 182, 196, 198, 203n, 211, 217
and determinism, 20
emotional effect of, 25
and microperceptions, 40
neglect of, 18
causality, 14–15, 19–21, 23–6, 35–6, 40, 64, 203n
Christianity, 97, 100–2, 105, 112, 117, 216–17
climate change, 209–14, 216–19, 221; *see also* ecological crisis
Cobb, John, 211, 213
commodification, 87, 220
concreteness, 22, 58–61, 128, 177
conformation, 18–19, 23–5, 34–5, 48, 91
Confucianism, 108, 112–19, 121n
Connolly, William, 216–21
consciousness, 3, 26–7, 29, 40–2, 126, 129–30, 132, 134–5, 139, 160–2
continental philosophy, 4
Conway, Anne, 106n
cosmopolitics, 189–90, 201n, 221
creativity, 3, 61, 70, 73n, 91, 96, 104, 172, 175, 220; *see also* novelty
creature, 96–102, 104–5, 105n, 106n
cultural anthropology, 4
cultural codes, 86
cultural theory, 4, 89, 91
culture, 8, 61, 86–7, 89, 97, 170, 175, 178

Datson, Lorraine, 202–3n
Deacon, Terrence, 4
Deleuze, Gilles, 14, 27, 30–1, 38–52, 54n, 61–2, 161–4, 167n, 168n, 172, 189, 201n, 202n
Derrida, Jacques, 53n, 64, 68, 75n, 211
Descartes, René, 13, 147
deterritorialisation, 148–9, 160–2, 165–6

231

Dickens: Charles, 98
difference, 26, 39–41, 43–5, 47, 50–2, 54n
Difference and Repetition, 47, 50–1, 54n
Douglas, Mary, 4
dreams, 125–6, 138, 140
dualism, 39, 45, 81
Dueck, Alvin, 134
Durkheim, Émile, 4

ecological civilisation, 215–16
ecological crisis, 170, 172, 181, 211–12
ecology, 3, 4, 5, 56, 172, 192, 202n, 213–14
economics, 218
education, 14, 118–19, 121n, 191–3, 209, 221
ego, 132, 139
emotion, 19, 25, 110, 120n, 136–7, 195, 214; *see also* feeling
empiricism, 17, 178
environmental crisis *see* ecological crisis
epistemology, 13, 23–4, 39, 111, 174, 184n, 202–3n
error, 2–3, 8, 13–14, 19, 22–3, 33, 65, 74n, 155, 160, 165–6, 173, 175, 179, 196
Esposito, Roberto, 202n
eternal object, 1, 73n, 128–9, 135–6, 138–40, 197; *see also* possibility; potentiality
eternal return, 50, 52–3, 54n
ethology, 149–50; *see also* animal thought
Evans, Gareth, 150
events, 2–3, 16, 42, 48, 52, 59
 connectedness of, 14, 34
 see also actual entity
evil, 172, 180–1; *see also* perpetual perishing
experience, 3, 15–16, 19, 24, 29, 32, 34, 36–8, 40, 43–4, 48, 51, 83–5, 92, 110, 117
 act of, 20
 delusive, 14, 22
 direct, 3, 19, 22, 24, 32–4, 45, 86, 211
 infallibility of, 22
 intensity of, 72, 127, 129–30, 133–4, 139, 180
 reliability of, 14
 temporality of, 22–3
 see also perception

Faber, Roland, 222
fallacy of misplaced concreteness, 22, 59, 88, 177, 191, 208, 212, 218; *see also* abstraction
family, 108, 116, 118–19, 122n

feeling, 26–7, 36, 90–3, 104, 126, 130, 137, 139, 173, 197, 214
 and experience, 91
 see also emotion
filial piety, 113, 118
Foucault, Michel, 88
freedom, 65, 109–10, 112, 128, 179, 196
Freud, Sigmund, 135

Galison, Peter, 202–3n
Garcia, Tristan, 20
gender, 98–9
gentleness, 72
God, 1, 56, 70, 96, 98–104, 126, 128, 131, 133–40, 182, 215
graded envisagement, 2–3
Griffin, David Ray, 9n, 26, 127
Guattari, Felix, 27, 43, 161–4, 167n, 201n

habit, 21, 49–50, 161–3, 165, 177–8, 180
Halewood, Michael, 203n
Hansen, Mark B. N., 26
Hardt, Michael, 188, 200n
Harman, Graham, 17, 19–20
Hartshorne, Charles, 167n
heart-mind, 113–16, 119, 121n
Heidegger, Martin, 33
Heraclitus, 52
Hesiod, 217
hierarchy, 62, 92–3
high-grade organisms, 3, 18, 31–2, 35, 40, 68, 111–12, 129–30, 153, 173, 176, 179, 181
Hosinski, Thomas, 120n, 129–31
humanism, 81, 195
humanity, virtue of *see* ren
Hume, David, 14–16, 20–1, 67, 111, 162–3, 203n
Husserl, Edmund, 17

ideality, 3, 8
ideology, 87–90, 93
Idiot, 187, 189–90, 194, 199–200, 201n, 202n
inhumanity, 85–7, 91, 93–4
initial aim, 128, 131, 134, 139; *see also* God
instinct, 109, 117, 135–6, 148, 154–5, 159–60, 162, 166
Irigaray, Luce, 63, 68

James, William, 37, 38, 48
Joseon Dynasty, 112–13, 117
Jung, Carl, 124–7, 132–40

Kant, Immanuel, 14–16, 21, 24, 34–5, 189, 203n
Keller, Catherine, 201n, 202n
khora, 60, 69
Kongzi *see* Confucianism
Kraus, Elizabeth, 136
Kristeva, Julia, 68

Lacan, Jacques, 4, 195
Lachmann, Rolf, 120n
language, 1, 5, 31, 36, 81–2, 147–9, 151, 156, 158–60, 165, 174, 195, 214
Latour, Bruno, 171
Lee, Bernard, 127–8
Leibniz, Gottfried, 41–3
Leroi-Gourhan, André, 165
Lewis, David, 15, 23
life, 26–7, 109–10, 147–8, 154, 164–5, 184
  fragility of, 58, 181, 220
limitation, 43, 46
Locke, John, 112
love, 99, 139, 170, 183
low-grade organisms, 3, 25, 109, 128–9, 173, 179
lure for feeling, 14, 127–9, 137, 187, 197, 199, 222

manipulation, 170–2, 179
Marx, Karl, 87–90, 92–3, 200–1n
mathematics, 31, 82, 137
Maxwell, Grant, 135
meaning, 32–3, 36–7, 64–5, 81, 83–6, 90, 93, 110–11, 119, 120n, 128, 171, 173, 175–6, 178, 180, 184n, 194–5, 197, 199
Meillassoux, Quentin, 15, 23–4, 53n
memory, 49–51, 133, 135
Mencius *see* Mengzi
Mengzi, 114–15, 121n
mental pole, 29, 104, 129, 134, 138–9
Metzinger, Thomas, 14
Middle Ages, 82
Modes of Thought, 103, 120n, 190
morality, 100, 104
mutual immanence, 30, 34, 63, 69–70
mutual transcendence, 70–1
myth, 86, 125

nature *see xing*
Negri, Antonio, 188, 200n
Neo-Confucianism, 112, 115, 121n, 122n
novelty, 3, 61, 128, 130, 139, 172; *see also* creativity

objectification, 20, 33, 35, 60–2, 68

objective immortality, 129
Odin, Steve, 135–6
ontology, 13, 24, 108, 112, 115
originary symbolism, 30–1, 34, 38–9, 41, 44–7, 49–50, 53n, 54n, 82

panexperientialism, 9n, 26, 167n; *see also* panpsychism
panpsychism, 9n, 26, 29, 148–9, 152, 155, 165, 167n; *see also* vitalism
peace, 183
perception, 1–4, 17–18, 24–5, 29–34, 39–46, 51, 59, 62, 82, 110, 125, 151, 156, 160, 166, 171, 173, 176, 178, 194–6, 203n, 209–10, 212, 214
  direct, 17–22, 24–6
  human, 4, 8, 25–6, 36, 43–5, 65, 71–2, 81–7, 93–4, 111–12, 124, 126–7, 130–1, 133–6, 147, 175, 178–81
  infallibility of, 3, 196
  modes of, 16, 18, 32, 34–6, 40, 47, 62–3, 67–8, 110, 130–1, 138–9, 152–3, 173–4, 177, 196
  non-human, 4, 40, 43, 46, 84, 106n, 147, 151, 179
  non-sensuous, 26, 125–6
  primacy of, 32
  reliability of, 14
  see also experience
perpetual perishing, 30, 50, 172, 180–1, 184
phenomenology, 18
physical pole, 29, 104, 138
Plotinus, 163, 167n
political philosophy, 56, 216
political system, 166
politicisation, 209–10, 212–14
politics, 96, 159, 189, 209–10, 213, 218, 220–2
possibility, 23, 62, 73n, 127–8, 131, 133, 136; *see also* eternal object; potentiality
potentiality, 23, 62, 73n, 91–2, 176, 197; *see also* eternal object; possibility
power, 53n, 88, 100–1, 110
prehension, 1, 2, 5, 39, 45, 120n, 128–9, 138
presentational immediacy, 2, 16–19, 21–2, 34–6, 59, 62–4, 66, 68, 110–11, 120n, 130–1, 138–9, 152–3, 155–6, 159–61, 174–9, 182, 196, 211
  and macroperceptions, 40
process, 17, 25–6, 30, 40, 45–7, 50, 52, 53n, 54n, 60, 70, 89–92, 127
Process and Reality, 1–2, 21, 23, 37, 45–6, 74n, 111, 133, 172, 175, 177, 180, 182, 186, 197, 204n

professionalisation, 191–4
propositions, 129–30, 187, 197–200
psyche, 125, 131–5, 138–40
psychology, 132, 134, 149
purpose, 3, 109–10, 128, 137, 173

reason, 71–2, 98–9, 130, 157, 165, 213–15
recombinable representations, 156–8
relationality, 2, 15, 24, 30, 32, 35–7, 43, 47, 60–1, 66, 108, 111–18, 152, 159, 182, 192, 223
religion, 1, 56, 58, 103–5, 181, 222
Religion in the Making, 103, 139
ren, 113–14, 116, 118
reterritorialisation, 148–9, 161–2, 165–6
ritual, 108, 112–13, 115–19
Robertson, Robin, 137
Romanticism, 97–8, 100–2, 104–5
Rorty, Richard, 60, 62
Russell, Bertrand, 9n, 13

sage, 116–18
Sartre, Jean-Paul, 160
science, 4, 177–8, 184n, 187, 189–90, 194, 199, 203n, 213, 218
    competitive, 187–90, 193–4, 198
    see also technoscience
Science and the Modern World, 2, 59, 100–1, 103, 137
scientific research, 177, 187–8
self, the, 93, 108, 111–13, 115–16, 118, 134
Sellars, Wilfrid, 151
sense-data, 14, 17–18, 20, 30, 35, 59, 111, 131, 155–6, 173–5, 196, 204n
sensitiveness, 70–2
Shaviro, Steven, 200n
Shelburne, Walter, 137
Shelley, Mary, 99, 102
Shelley, Percy, 99–100, 102
shengren *see* sage
simple location, 2, 15, 30, 59, 218
Slusser, Gerald, 135
social cohesion, 108–10, 113, 153
social environment, 85
social evolution, 8, 70
social experience, 90–2, 97, 159
social theory, 85, 88, 108
society
    development of, 3–5, 8
    human, 3–5, 8, 72, 85–6, 89–90, 108–9, 117–18, 127, 130, 170–1, 179–80, 186, 214–16, 218

sociology, 85–8
solipsism, 3, 18, 30, 34
space, 34, 59
Stengers, Isabelle, 19, 81, 187, 189–90, 193, 199, 201n, 202n, 203n, 204n
stimulus-independence, 148–50, 156–7, 159
subject of experience, 17, 33, 42, 44–5, 49, 108, 111, 130, 163–4, 179; *see also* subjectivity
subjectivity, 36, 61, 109, 148, 173–4, 177; *see also* subject of experience
symbolic reference, 3, 8, 18, 22, 35–6, 38, 40, 51, 62–3, 65, 68, 84, 111, 120n, 130–1, 138, 153, 158–61, 166, 171, 173–5, 179, 194–6, 211; *see also* error
synthesis, 38, 44, 47–51, 59, 64–5, 67–8, 152–3, 155, 162–4, 174

technoscience, 186–7, 192–5, 198; *see also* science
Thacker, Eugene, 27
theology, 64
    political, 96–102, 104
Tillich, Paul, 131
time, 22–3, 30, 34–5, 37–8, 46–52, 218
truth, 3, 13–14, 65, 157–8, 198
    progress in, 8

Uexküll, Jakob von, 154–5
Ulanov, Ann Belford, 134
unconscious, 132–4, 138–9
    collective, 132–3, 135–6

value, 81, 97, 103–4, 126, 128, 194
Van Dyck, Barbara, 187–90, 194, 198–9
Virno, Paulo, 201n
vitalism, 26, 27, 147–9, 153–4, 159, 162–4; *see also* panpsychism
von Franz, Marie-Louise, 138

Western philosophy, 3, 13–14, 23, 98
Williams, Raymond, 89–93
Wilson, Jessica, 168n
Wollstonecraft, Mary, 98–9, 102
Woolf, Leonard, 9n
Woolf, Virginia, 9n

xiao *see* filial piety
xin *see* heart-mind
xing, 114–15

Zeno, 38, 46

EU representative:
Easy Access System Europe
Mustamäe tee 50, 10621 Tallinn, Estonia
Gpsr.requests@easproject.com

www.ingramcontent.com/pod-product-compliance
Lightning Source LLC
Chambersburg PA
CBHW051115230426
43667CB00014B/2592